Tobias Laschütza

# Multiscale modelling of crazing in glassy polymers under cyclic loading

**Schriftenreihe des Instituts für Mechanik**
**Karlsruher Institut für Technologie (KIT)**

**Band 12**

Herausgeber:
Prof. Dr.-Ing. habil. Peter Betsch
Prof. Dr.-Ing. habil. Thomas Seelig

Eine Übersicht aller bisher in dieser Schriftenreihe erschienenen Bände finden Sie am Ende des Buches.

# Multiscale modelling of crazing in glassy polymers under cyclic loading

by
Tobias Laschütza

KIT Scientific Publishing

Karlsruher Institut für Technologie
Institut für Mechanik

Multiscale modelling of crazing in glassy polymers under cyclic loading

Zur Erlangung des akademischen Grades eines Doktor-Ingenieurs
von der KIT-Fakultät für Bauingenieur-, Geo- und Umweltwissenschaften
des Karlsruher Instituts für Technologie (KIT) genehmigte Dissertation

von Tobias Laschütza, M.Sc.

Tag der mündlichen Prüfung: 19. November 2024
Hauptreferent: Prof. Dr.-Ing. habil. Thomas Seelig
Korreferent: PD Dr.-Ing. habil. Sebastian Pfaller

**Impressum**

KIT Scientific Publishing

Karlsruher Institut für Technologie (KIT)
KIT Scientific Publishing
Straße am Forum 2
D-76131 Karlsruhe

KIT Scientific Publishing is a registered trademark
of Karlsruhe Institute of Technology.
Reprint using the book cover is not allowed.

www.bibliothek.kit.edu/ksp.php | E-Mail: info@ksp.kit.edu | Shop: www.ksp.kit.edu



Print on Demand 2025 – Gedruckt auf FSC-zertifiziertem Papier

ISSN 2363-4936
ISBN 978-3-7315-1409-1
DOI 10.5445/KSP/1000178410

# Abstract

*Crazing* is the primary damage mechanism in amorphous thermoplastics. Despite decades of extensive research, craze models that allow the study of fracture processes involving crazing and shear yielding under cyclic loading have yet to emerge. Crazing describes the localised formation and dilatant growth of crack-like defects, consisting of fibrillated matter interspersed with voids. This network of highly oriented fibrils is of significant practical importance, as it enables load transfer between its surfaces, resulting in a substantial increase in the material's fracture toughness.

This cumulative dissertation aims to elucidate the cyclic crazing response and its interaction with the adjacent bulk material by utilizing a multiscale approach that combines *molecular dynamics* and *continuum micromechanics.* To this end, a micromechanics-inspired constitutive model for crazing is developed, enhanced with results from molecular dynamics simulations and employed as a traction separation law along the ligament of a mode I crack propagation problem. In a finite strain setting, the model accounts for (i) the morphological changes from microvoids to mature fibrils during craze initiation, (ii) craze thickening due to viscoelastic deformation of existing craze fibrils, (iii) craze thickening due to the transient process of viscoplastic conversion of bulk material into new fibrillated craze matter and (iv) premature jamming of the craze fibrils during unloading. The structural behaviour of the craze fibrils is modelled as string-like. This leads to creep recovery of the fibril deformation when they are stress-free, which facilitates continuous fibril drawing across loading cycles. As the craze is taken to break down at a critical thickness, the current craze fibril length can be considered a measure of cyclic damage accumulation.

To enrich the polymer physical understanding of the involved processes, molecular dynamics simulations of a generic bead-spring model are conducted. The mechanical response of sole fibrillated craze matter and the bulk-craze interaction in glassy polymers under cyclic loading is analysed. The results support the string-like representation of craze fibrils as they undergo a stress-free folding motion during unloading. Further essential findings of this study are the complex stress response and the driving mechanisms leading to the hysteresis. Extensive postprocessing of the simulation results enables the incorporation of mechanisms and experimentally inaccessible material parameter scopes into the crazing model through a *bottom-up* approach.

The molecular dynamics informed continuum model is employed in a finite element study to analyse mode I craze and crack growth under cyclic loading in brittle and ductile glassy polymers. The model, along with the calibration methodology, successfully reproduces important craze, shear band and failure characteristics reported in the experimental literature. These include, for instance, the craze contour, the initiation of shear bands at the current crack tip that arch towards the craze, which shares much resemblance to the so-called *epsilon-shaped deformation zone* and properties of *normal fatigue crack propagation*. A novel insight form this study is the delayed crack propagation due to plasticity-induced unloading of the craze at the crack tip.

# Kurzfassung

*Crazing* ist der primäre Schadensmechanismus in amorphen Thermoplasten. Trotz umfangreicher Forschung in den letzten Jahrzehnten existieren bislang keine Crazing-Modelle, die eine detaillierte Untersuchung von Bruchprozessen in Wechselwirkung mit Scherfließen unter zyklischer Belastung ermöglichen. Crazing beschreibt die lokalisierte Bildung und dilatante Ausdehnung rissähnlicher Defekte, die aus fibrillierter Materie bestehen und mit Poren durchzogen sind. Dieses Netzwerk hochorientierter Fibrillen ist von hoher praktischer Bedeutung, da es einen Lasttransfer zwischen den Crazeflächen ermöglicht und somit signifikant die Bruchzähigkeit des Materials erhöht.

In dieser kumulativen Dissertation wird die Wechselwirkung zwischen Crazing und dem umgebenden Grundmaterial des glasartigen Polymers durch einen Multiskalenansatz mittels *Molekulardynamik* und *Kontinuumsmikromechanik* unter zyklischer Belastung untersucht. Hierfür wird ein mikromechanisch inspiriertes konstitutives Modell für Crazing entwickelt und durch molekulardynamische Simulationsergebnisse erweitert, um es anschließend als Kohäsionszone entlang des Ligaments in einem Modus I Rissproblem einzusetzen. Im Rahmen finiter Deformationen berücksichtigt das Materialmodell (i) die morphologischen Veränderungen von Mikroporen zu vollständig entwickelten Fibrillen während des Initiierungsprozesses, (ii) die Dickenänderung des Crazes infolge viskoelastischer Deformationen bestehender Crazefibrillen, (iii) die Dickenänderung des Crazes durch den transienten Prozess der viskoplastischen Umwandlung von Grundmaterial in neues fibrilliertes Crazematerial sowie (iv) ein frühzeitiges Verklemmen der Fibrillen während der Entlastung. Das Strukturverhalten der Crazefibrillen wird als seilartig idealisiert, was zur Kriecherholung der Fibrillenverformung führt, während die Fibrillen spannungsfrei sind. Dies begünstigt ein kontinuierliches Fibrillenziehen über mehrere Lastzyklen hinweg. Der Craze versagt bei einer kritischen Dicke, sodass die aktuelle Fibrillenlänge ein Maß für die zyklische Schadensakkumulation darstellt.

Zur Verbesserung des polymerphysikalischen Verständnisses der Prozesse werden molekulardynamische Simulationen eines generischen Bead-Spring-Modells durchgeführt. Dabei wird das mechanische Verhalten eines Crazes sowie die Interaktion mit dem Grundmaterial unter zyklischer Belastung analysiert. Die Simulationsergebnisse bestätigen das seilartige Verhalten der Fibrillen, da diese während der Entlastung eine spannungsfreie Faltbewegung durchlaufen. Weitere zentrale Ergebnisse sind die komplexe Spannungsantwort und die zugrundeliegenden Mechanismen der Hysterese. Umfangreiche Auswertungen der Simulationen ermöglichen die Einbindung von Mechanismen und experimentell unzugänglichen Materialparametern in das Crazing-Modell durch einen *Bottom-Up*-Ansatz.

Das hierdurch molekulardynamisch erweiterte Kontinuumsmodell wird in einer Finite-Elemente-Studie eingesetzt, um die Craze- und Rissausbreitung im Modus I unter zyklischer Belastung in spröden und duktilen amorphen Polymeren zu analysieren. Das Modell sowie die Kalibrierungsmethodik reproduzieren erfolgreich wichtige Crazing-, Scherband- und Versagenscharakteristiken aus der experimentellen Literatur. Dazu gehören beispielsweise die Crazekontur, die Initiierung von Scherbändern an der aktuellen Rissspitze, die sich zum Craze hin neigen und ausgeprägte Ähnlichkeit mit der sogenannten *epsilon-förmigen Deformationszone* aufweisen, sowie Charakteristiken der *normalen Ermüdungsrissausbreitung*. Eine neuartige Erkenntnis dieser Studie ist die verzögerte Rissausbreitung aufgrund der plastizitätsinduzierten Entlastung des Crazes an der Rissspitze.

# Acknowledgments

This PhD thesis was written during my time as a research associate and doctoral candidate at the Institute of Mechanics (IFM) of the Karlsruhe Institute of Technology (KIT), within the research group of Prof. Thomas Seelig. He introduced me to the fascinating fields of fracture mechanics and material modeling of polymers. This thesis would not have been possible without his exceptional mentorship and support throughout my scientific journey. I am deeply grateful for his extensive engagement in my research, the countless hours of insightful on-topic (and off-topic) discussions and the freedom and encouragement to pursue my ideas, including the opportunity for a research stay at UBC. Thank you for your guidance, inspiration and dedication.

My special thanks go to PD Sebastian Pfaller for serving as the second examiner for my thesis and for his genuine interest in my work. Furthermore, I am sincerely grateful to Prof. Jörg Rottler, who introduced me to the exciting world of molecular dynamics simulations during my six-month research stay at the Stewart Blusson Quantum Matter Institute at UBC Vancouver, which was partially funded by KHYS. The inspiring atmosphere and many fruitful discussions greatly enriched my understanding and made a significant contribution to this research. Thank you for your genuine interest in my work and your remarkable patience with my engineering approaches.

Moreover, at IFM and UBC, I greatly benefited from numerous colleagues through insightful on- and off-topic discussions, their efforts in fostering an excellent working atmosphere and many memorable events – ranging from work-related conferences to various social gathering and sports activities. Thank you all sincerely for the wonderful experience and the fond memories.

Finally, I would like to thank my family, friends, and most importantly, Sina, for their continuous support and unconditional love.

Karlsruhe, February 2025 Tobias Laschütza

# Contents

# 1. Introduction

## 1.1. Motivation

Polymers are an important class of engineering materials. Research advancements and the combination of outstanding mechanical, thermal, electrical, optical and manufacturing properties as well as their biocompatibility have popularised polymers across a wide range of applications from simple packaging to high-performance elements across essentially all industries such as automotive, aerospace, electronics and biomedical devices [25]. From an engineering perspective, understanding the mechanical response of polymeric materials under a broad range of loading conditions is critical to ensuring their proper functioning. Cyclic (fatigue) loading is particularly crucial, as many engineering materials spend part or even their entire lifetime under these conditions.

Amorphous thermoplastics are an important subset of engineering polymers. They are characterised by a disordered molecular structure in which molecular chains form an entangled network without chemical cross-links. With a glass transition temperature above room temperature, these materials exhibit high stiffness in their glassy state at room temperature. As the temperature approaches and surpasses the glass transition point, their ductility increases and they eventually transition into a rubbery state. The key damage mechanism in glassy thermoplastics is crazing, which describes the dilatant process of localized formation and growth of crack-like defects, cf. [34, 52, 53]. Crazes are a network of highly oriented fibrils interspersed with voids as depicted in Figure 1.1. Oriented in the direction of maximum principal stress, the main fibrils are several tens of nanometre thick and can grow up to a few micrometres in length, whereas the craze length can

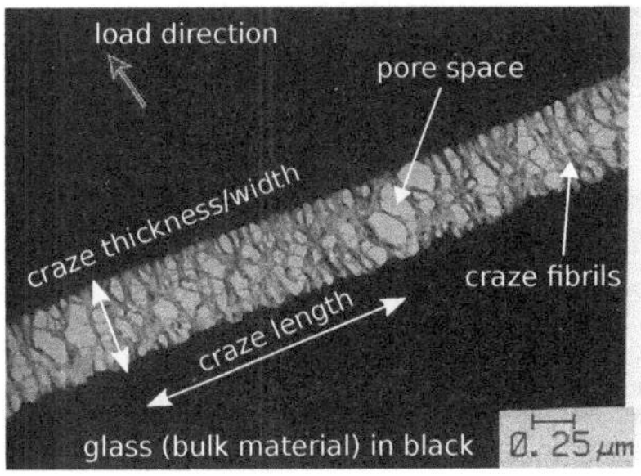


**Figure 1.1.:** Craze in polystyrene from [62] with additional annotations.

extend over several millimetres (cf. arrow and length scale in Figure 1.1). For sufficiently high entanglement densities, cf. [77], the main fibrils are commonly interconnected by so-called cross-tie fibrils. Unlike cracks, the fibrillated craze matter enables a significant load transfer between its surfaces, which substantially increases the material's fracture toughness. This gives the crazing process significant practical importance. Simplified, the crazing process involves three stages: First, *craze initiation*, which is marked by the formation of microvoids that eventually coalesce into an interconnected void space. Second, *craze growth in thickness*, which occurs in the direction of maximum principal stress as surrounding bulk material is drawn in from the so-called *active zone*. At the molecular level, this process is understood to be driven by chain scission and disentanglement. Finally, *craze breakdown* as the fibrils rupture, which is often observed when the craze thickness approaches a critical width.

Despite the intense experimental research on crazing over the past decades, cf. reviews in [34, 46, 48, 49, 66], theoretical-computational studies involving continuum modelling [12, 26, 27, 82, 84, 96, 97] and molecular dynamics simulations [11, 32, 60, 76, 100, 102] are primarily confined to monotonic loading conditions. Hence, the understanding of the governing micromechanisms during cyclic loading and cyclic damage accumulation remains incomplete and inconclusive. Compared to monotonic loading, modelling the response of the fibrillated craze matter under cyclic loading is more challenging, particularly due to the ambiguity in the structural response of craze fibrils during unloading and reloading. For instance, this includes the relaxation and creep recovery of the craze fibrils when fibril drawing is interrupted, the uncertainty in selecting an adequate fibril deformation model that encompasses the unloading and reloading response and the understanding of cyclic damage accumulation by, for instance, either fibril drawing and creep deformation [15] or fibril buckling [19, 64]. Moreover, a combination of experimental and theoretical evidence indicates the occurrence of compressive stress during unloading while the macroscopic deformation is still tensile [23]. As this behaviour scales with the craze length, it may be interpreted as a jamming of the craze fibrils arising from the dilatant nature of the crazing process and the low probability of fibrils perfectly aligning during unloading. Although it is a physically plausible interpretation, the consequences on the adjacent bulk material are not yet understood. Further challenges arise from the length scale involved in the crazing process, which makes it difficult to obtaining detailed mechanical insight through experimental investigations. This also poses challenges for theoretical-computational studies, as calibrating material parameters of constitutive models through the classic engineering approach using experiments is not readily possible.

While the previous challenges focused on modelling of a uniform craze, additional model complexity arises when studying the craze in the context of cyclic crack propagation. In this scenario, the craze interacts with the adjacent bulk and exhibits nonuniform deformation along its length. Moreover, fracture mechanisms change with the loading amplitude [90]. For instance, at low amplitudes, crack growth may not occur in every cycle, but rather *discontinuously* in jumps [50, 83]. Damage accumulates over hundreds of cycles as the craze length grows. Upon reaching a critical limit, fibrils at the crack tip rupture abruptly within a single cycle, resulting in a rapid crack advance and a shrinking

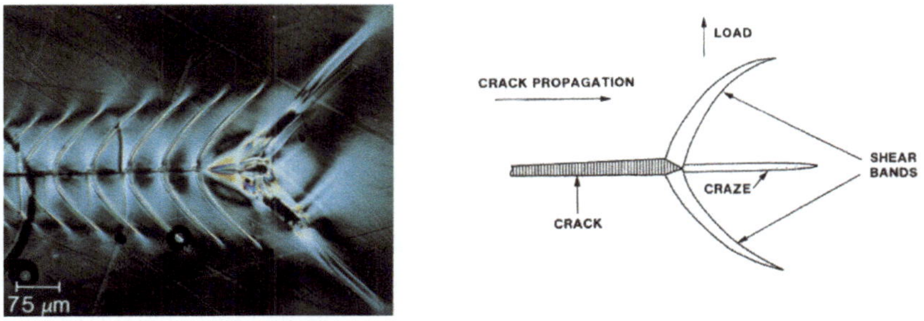


**Figure 1.2.:** Epsilon crack tip plastic zone in polycarbonate (left) with a schematic (right) from [94].

craze length, after which the damage cycle repeats. Könczöl et al. [50] reported a transition from mainly fibril drawing in the first half of the craze's lifetime to predominantly fibril creep deformation in the second half leading up to the crack jump. This type of crack propagation is referred to as *discontinuous* or *retarded* fatigue crack growth, whereas *normal* fatigue crack growth involves crack propagation in every cycle. An increase in loading amplitude may give rise to the formation of regularly spaced shear bands in ductile polymers, which initiate at the crack tip and arch in the direction of crack propagation as shown in Figure 1.2. The crack growth is again discontinuous and the combination of the two shear bands with the craze ahead of the crack tip leads to the name-giving fracture process regarded as *epsilon discontinuous crack growth* and the local description of the crack tip as *epsilon crack tip plastic zone* [63, 92, 94]. From a fracture mechanics perspective, this is noteworthy, since both, shear bands and the craze, initiate from the crack tip. In contrast, craze initiation is often observed at the intersection of already formed shear bands, e.g., at blunt notches [66] or microvoids [105]. Besides the fracture mechanical aspect, the interaction between shear yielding and crazing is also of technical relevance, since the formation of shear bands increases the load bearing capacity by shielding the craze from further damage [94].

Concluding, although crazing holds significant practical importance in polymer engineering, theoretical models that allow the study of fracture processes involving crazing and shear yielding under cyclic loading have yet to emerge.

## 1.2. Aim and methods

The aim of this work is to elucidate the cyclic crazing response and its interaction with the adjacent bulk material under mode I craze and crack growth, as, for instance, shown in Figure 1.2. A physically motivated crazing model that realistically captures the response under cyclic loading is not yet available in the literature. This leads to the following three challenges and solution approaches involving the highlighted *methods*:

**Craze model**
The first challenge concerns the development of an adequate craze model within a suitable framework. The most prominent existing craze models were developed by Van der Giessen and co-workers [26, 96, 97] and Boyce and co-workers [82, 84]. However, both consider only monotonic loading conditions, thereby omitting micromechanisms that become important under cyclic loading. The key obstacles in model development can be summarised as follows: The length scale involved in the crazing process restricts extensive experimental evaluations. This renders a purely phenomenological model with generic material parameters unsuitable. Moreover, the crazing process involves complex micromechanisms, such as craze thickening as transient process of bulk conversion into fibrillated craze matter with lower density or the morphological transition from thick to thin (mature) fibrils during craze initiation. Besides the micromechanisms, realistic constitutive descriptions are necessary to represent the deformation behaviour of both bulk material and craze fibrils during unloading and reloading, as well as processes such as craze initiation, fibril drawing and fibril breakdown.

A *physically motivated continuum-micromechanical model* (see Chapter 2) can address these challenges effectively. It enables the separation of bulk and craze responses, which facilitates adequate model selection and provides the flexibility to readily replace model components as new knowledge emerges. It further allows access to important micromechanical quantities, such as the fibril stress. Moreover, it simplifies the accommodation of important micromechanisms and permits the association of experimentally measurable parameters with them. Altogether, these features make sensible parameter scopes more naturally accessible, which remedies the lack of experimental knowledge. Combined with a *material parameter normalisation*, the model response can be systematically investigated. The overall aim to study the interaction with the adjacent bulk material, e.g., undergoing shear yielding under mode I crack growth makes the continuum-mechanical framework a suitable choice. The craze model can be applied as traction separation law along the mode I symmetry plane. The model development and testing can be efficiently conducted using Python, as it offers capabilities of automatic differentiation and includes solvers for nonlinear system of equations.

**Uncertainties in molecular mechanisms**
The second challenge pertains to the uncertainty in molecular mechanisms involved in the cyclic deformation of craze matter. The constitutive craze model is inherently based on simplification and assumptions that stem partially from the incomplete knowledge of the cyclic crazing process. This can be divided into two parts: On the one hand, there is an inconclusive polymer physical understanding regarding the structural behaviour of crazes during unloading and reloading. This poses an significant issue, since it impacts the picture of the damage mechanism during cyclic (fatigue) loading. On the other hand, there is a lack in detailed knowledge concerning the mechanical response of craze deformation across a broad range of loading conditions.

Considering the length scale, *molecular dynamics simulation* (see Chapter 3) provides a potential solution. It yields insight into many molecular scale mechanisms of crazing in particular the ones concerning the entanglement network. Thus, they contribute to establishing the microscopic picture which is often assumed in micromechanical models, yet difficult to evaluate experimentally. Moreover, the controlled conditions in the simulations enables a systematic bottom-up transfer of results to the continuum model, as molecular dynamics simulations and the micromechanical model complement each other effectively. The Large-scale Atomic/Molecular Massively Parallel Simulator (LAMMPS) package [69, 95] offers an efficient way to conduct the molecular dynamics simulations.

**Mode I crack growth**
The third challenge is the analysis of craze and crack growth under mode I loading. In addition to the craze model, this demands a realistic constitutive description of the bulk material and the solution of the boundary value problem accounting for both materials. Furthermore, the material parameters of both, bulk and craze model, need to be adequately calibrated to yield meaningful results. Finally, the quality of the model needs to be assessed.

Mode I crack growth can be studied in a physically meaningful way by merging the previous results into a *multiscale model with normalised material parameters* and qualitatively comparing simulation results to a wide range of experimental observations for brittle and ductile polymers reported in the literature (see Chapter 4). The molecular dynamics analysis allows for the incorporation of relevant mechanisms as well as representative parameter scopes, which cannot be deduced from experiments, into the craze model. The cyclic viscoplastic finite strain deformation of the adjacent bulk material can be modelled with an established and experimentally calibrated glassy polymer model. Adequate *normalisation of boundary value problem dimensions* mitigates size effects. Combined with the material parameter normalisation of bulk and craze models, this enables a systematic evaluation of generic, yet representative glassy polymer responses. The commercial *finite element* software ABAQUS [2] allows for an efficient solution of the boundary value problem by incorporating the non-standard materials via user subroutines written in FORTRAN. *Dual number automatic differentiation* is a suitable tool for the efficient implementation of the material models in FORTRAN.

## 1.3. Outline

This cumulative dissertation is organised into five chapters, where chapter 2 to 4 contain the stand-alone publications. As they can be read independently, a certain degree of repetition occurs between the chapters.

**Chapter 2 reproduces [56].** In this chapter, the continuum micromechanical crazing model for cyclic loading is presented. The micromechanical finite strain setting of the craze element, comprising bulk and fibrillated craze matter, is introduced in Section 2.2. The constitutive equations governing the different components and mechanisms within the model accompanied by a detail motivation and the efficient numerical realisation are addressed in Section 2.3. The influence of the parameters, with emphasis on the viscosities associated with fibril drawing and fibril creep deformation, on the craze element response under uniaxial deformation is analysed in Section 2.4.

**Chapter 3 reproduces [55].** In this chapter, the cyclic craze response within molecular simulations of a coarse-grained bead spring model is studied from a polymer physics points of view. The model setup and the methods are outlined in Section 3.2. The molecular mechanisms leading to the macroscopic cyclic stress-strain response of a sole craze and for coexisting fibrillated craze and bulk matter are analysed in Section 3.3. Links to animations and supplementary simulation results are presented in the Appendices 3.A and 3.B, respectively. Appendix 3.C presents additional, unpublished results on semiflexible polymer glass, which extends the analysis in [55] and addresses the role of the bending stiffness on the macroscopic stress-strain response.

**Chapter 4 reproduces [57].** In this chapter, craze and crack propagation under cyclic mode I loading for generic brittle and ductile glassy polymers is analysed. The craze model introduced in Chapter 2 is enhanced by molecular dynamics results from Chapter 3, leading to a molecular dynamics informed continuum model for crazing. A summary of the crazing model including remarks on the parameter calibration and efficient numerical implementation in Abaqus via dual number automatic differentiation is given in Section 4.2. Details on the molecular dynamics based parameter calibration are provided in Appendix 4.A. Finite element simulations of the mode I boundary value problem are subject of Sections 4.3 and 4.4. Section 4.3 presents the boundary value problem and the analysis of brittle glassy polymers. The interaction of crazing and shear yielding in ductile glassy polymers is studied in Section 4.4. The constitutive model to describe finite strain viscoplastic shear yielding in the bulk material is briefly summarised in Appendix 4.B.

**Chapter 5** concludes this work with a summary and an outlook for potential further research.

In this cumulative dissertation, introductions to the theoretical foundations are omitted. Instead, the reader is referred to the primary textbooks used in this work for nonlinear continuum mechanics [40], finite element method and computational plasticity [13, 85], polymer mechanics [14], polymer physics [88], and molecular dynamics [30]. Throughout this thesis, standard notation is employed, using boldface symbols to represent vectors and second-order tensors, as well as Cartesian index notation for their components. Also note, minor editorial adjustments to the original publications have been made to ensure a more consistent format for this dissertation.

# 2. A continuum-micromechanical model for crazing in glassy polymers under cyclic loading

**This chapter reproduces:***
Laschuetza T and Seelig T. "A continuum-micromechanical model for crazing in glassy polymers under cyclic loading". In: *Mechanics of Materials.* 189: 104901, 2024. DOI: 10.1016/j.mechmat.2023.104901

---

**Abstract:** A micromechanics-inspired constitutive model is developed to describe the deformation behaviour of fibrillated material within crazes in glassy polymers subjected to cyclic loading. In a finite strain setting, the model accounts for the morphology change taking place by the drawing of material from the intact bulk polymer into craze fibrils and their transition from primitive to mature fibrils. Building on previous research, fibril drawing is described as a viscoplastic process. A novel contribution of this study is the incorporation of viscoelastic deformation of existing fibrils, which is motivated by experimental observations. This new perspective allows for creep recovery, especially during the unloading phases of cyclic deformation.

A parameter study which pays special attention to the role of the characteristic times scales of fibril drawing and fibril creep in relation to the imposed loading rate illustrates the performance of the model. Since the model is designed as an input to cohesive fracture simulations in glassy polymers, its response under monotonic loading is analysed and compared to existing crazing models. Of primary interest, however, is the model behaviour under cyclic loading which is investigated for different loading scenarios up to fibril failure. The study highlights the complex interplay between the two viscous mechanisms and how they influence the local deformation behaviour of the craze matter as well as the number of cycles until failure.

**Keywords:** Craze, Cyclic loading, Micromechanical continuum model, Failure

---

## 2.1. Introduction

Crazing, i.e. the localized formation and growth of narrow zones of fibrillated matter interspersed with voids, is the key damage mechanism in glassy thermoplastic polymers. Crazes are typically oriented normal to the direction of maximum tensile stress and are in shape similar to cracks. Yet, contrary to cracks, crazes have a significant load carrying capacity owing to a multitude of thin fibrils of stretched polymer material which bridge the craze surfaces. The mechanical response of this fibrillated craze matter, hence, is of central interest in the behaviour of crazes. From intense studies over the past decades – see, e.g., the reviews in [34, 46, 49] – a reasonable understanding of the involved macromolecular and continuum-mechanical processes has emerged. For instance, it is nowadays well accepted that the formation and growth of crazes is governed by two deformation mechanisms: firstly, *drawing* of new material into fibrils at the craze/bulk interface and, secondly, *creep* of the existing fibrils. Anticipating a more thorough discussion in later chapters, it may be conjectured already here that the second mechanism becomes particular important during unloading phases of cyclic loading when the stress is too low to cause further pulling-in of new material and creep contraction of existing (eventually stress-free and loose hanging) fibrils is driven by the internal (back-)stress that originates from the highly stretched molecular chains.

Various theoretical-computational modelling approaches have been followed in order to gain a better insight into the micromechanics of crazing and also to incorporate the crazing mechanism in failure analyses (e.g. crack growth) on a larger length scale. In the latter case, and owing to the localized appearance of crazes, cohesive zone models for crazing – e.g. [26, 80, 97, 99] – have widely been utilized, but also continuum descriptions have been developed and employed, e.g. [33, 35, 84], for the simulation of crazing processes. On a smaller length scale the fibrillation process was studied by means of lattice-type models, e.g. [81], by continuum-mechanical models, e.g. [12, 36, 58], as well as by molecular dynamics simulations, e.g. [32, 76, 100]. Besides a better understanding of the crazing process itself along with its macromolecular implications such as disentanglement, another aim of these small-scale investigations was the determination of the (cohesive) craze response to be used in analyses on a larger scale, for instance as a traction-separation law in cohesive zone models. A key issue here is, and this brings us to the aim of the present work, that these computational studies – almost exclusively – consider crazing under monotonic loading.

Conversely, the craze response to cyclic loading, particularly focusing on fatigue crack propagation, has been extensively studied in experiments. These studies, from which only a few are mentioned here for brevity, have established that the highly transient and eventually interrupted crazing process under cyclic loading is of significant importance. For instance, in a study on several glassy polymers, Skibo et al. [83] observed that in the range of low loading amplitudes crack advance is likely to take place not in every cycle but *discontinuously* by jumps after remaining stationary for hundreds of fatigue cycles. Könczöl et al. [50] referred to this phenomenon as *retarded* fatigue crack propagation and reported a transition from mainly fibril drawing in the early stage to predominantly fibril creep deformation until jump-like further crack advance when a certain amount of damage by disentanglement might be reached. The issue of a certain disentanglement time of craze fibrils under cyclic loading was investigated by Schirrer et al. [79]. Aiming at studying the mechanical response of the craze matter, Pulos and Knauss [71] computed the normal stress distribution along a craze from the measured opening displacement profile during various stages of a cyclic loading-unloading process. Such a computation (performed by several other authors) relies on the numerical evaluation of an integral equation that governs the behaviour of the surrounding linear elastic medium. It has, however, been shown by Warren et al. [104] that this approach is ill-conditioned in the sense that small deviations in the input displacement profile give rise to significant changes in the computed stress. Interesting phenomena involving the interaction of crazes and shear bands during crack growth in glassy polymers under cyclic loading have been studied by Takemori [94] and co-workers. Direct measurements of the mechanical response of the fibrillated matter of existing crazes (i.e. no drawing in of new material) under cyclic loading were performed by Kambour and Kopp [47] and by Hoare and Hull [38]; both studies indicate a pronounced viscoelastic behaviour of the craze fibrils.

The aim of this study is to delve into the interplay between the drawing of bulk material into craze fibrils and the creep deformation of these fibrils under cyclic loading. By simplifying our approach – for instance, through neglecting the impact of cross-tie fibrils [18] – we examine the behaviour of a single fibril, treated as a *representative element*, which undergoes uniaxial tensile deformation after being drawn from the surrounding bulk material. In a finite strain setting, thereby we account for the transition from fibril formation into a *primitive fibril* and further into a *mature fibril* which corresponds to an experimentally observed draw ratio. Fibril drawing is described as a viscoplastic process while the (creep) deformation behaviour of the existing fibril is taken viscoelastic. The present work focuses on the model development and its analysis within a detailed parametric study and is meant as a first step. Subsequent work will utilize the model in a finite element framework to investigate the craze and crack growth in glassy polymer under cyclic loading while also accounting for shear yielding in the surrounding bulk material.

The present paper is organized as follows: In the following Section 2.2, the micromechanical model of the crazing process is developed and the coupling between the fibril behaviour and the macroscopic response is discussed in detail. The finite strain constitutive equations governing the different components and mechanisms within the model are presented in Section 2.3, which ends with a summary of the model and remarks concerning its computational evaluation. Section 2.4 contains a parametric study of the model, which is evaluated as a unit cell under uniaxial deformation. The study focuses primarily on the two dimensionless parameters characterising the key mechanisms of fibril drawing and fibril creep. For a deeper understanding of the effect of these parameters, we first look at the model behaviour under monotonic loading and thereafter analyse, as the major point of interest, the model response under cyclic loading up to failure. Conclusions concerning findings from this study and a critical discussion of the model results compared to experiments and previous simulations from the literature as well as an outlook towards future extensions and utilizations of the model are discussed in Section 2.5.

## 2.2. Micromechanical model

The model to be developed builds upon the micromechanical considerations depicted in Figure 2.1. Following Boyce and coworkers [82, 84], we consider an elementary volume of material referred to as a *micromechanical craze element* that undergoes a transition from uncrazed bulk to crazed material. Prior to craze initiation, the craze element consists solely of bulk material of an initial thickness $h_0$, the so-called primordial thickness. Upon craze initiation, the craze element comprises layers of fibrillated matter with the undeformed length $\xi_0$ and bulk material with the undeformed length $\chi_0$. As fibril drawing progresses, the fibril length $\xi_0$ grows while the bulk length $\chi_0$ shrinks. As indicated in Figure 2.1 and following the widely used modelling assumptions (e.g. [53]), the complex craze microstructure is idealised by neglecting cross-tie fibrils. That is, only string-like fibrils in the direction of the maximum principal stress, defining the coordinate system in

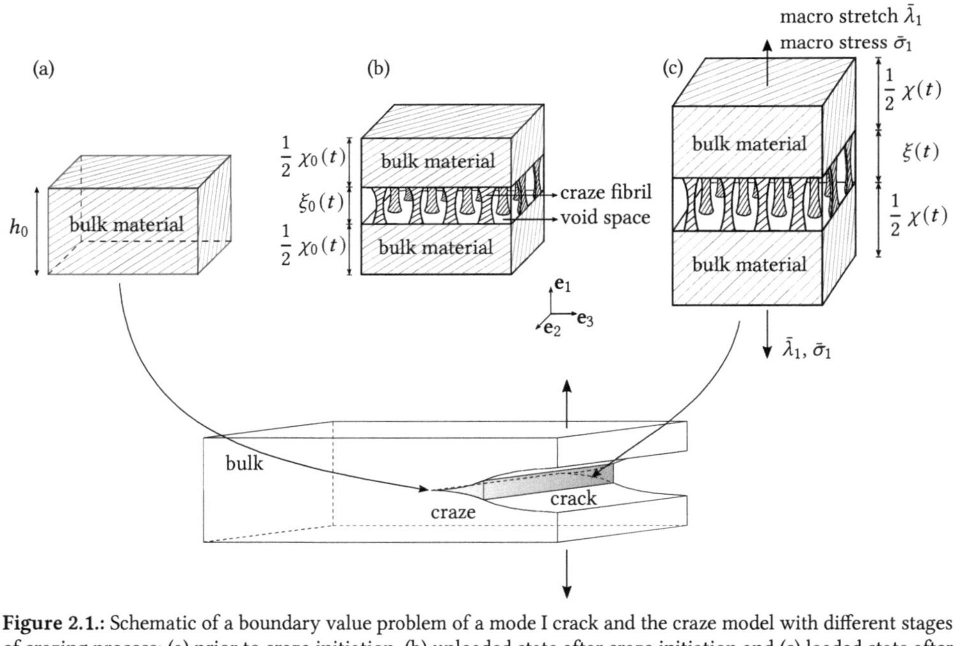


**Figure 2.1.:** Schematic of a boundary value problem of a mode I crack and the craze model with different stages of crazing process; (a) prior to craze initiation, (b) unloaded state after craze initiation and (c) loaded state after craze initiation.

Figure 2.1, are considered. Based on this simplification and adhering to the principle of mass conservation, the thicknesses of the two phases are related by

$$h_0 = \chi_0 + \frac{\xi_0}{\lambda_c}, \tag{2.1}$$

where $\lambda_c$ denotes the fibril extension ratio, defined here as the ratio of bulk density $\rho_b$ to the density $\rho_c$ of the fibrillated craze matter:

$$\lambda_c = \frac{\rho_b}{\rho_c}. \tag{2.2}$$

Note, that $\lambda_c$ here describes the maximum density ratio attained during the loading history and is with respect to the unloaded configuration. This is in some contrast to the extension ratio determined from experimental measurements (e.g. [52, 53]) or from molecular dynamics simulations (e.g. [76, 102]) quantifying typically the density ratio in the loaded state. However, the definition of $\lambda_c$ in the unstressed state considered here appears more practical.

### 2.2.1. Morphology change

The crazing process is understood to start from the formation of micro-voids in the bulk material and comprises a topological transition from isolated voids to an interconnected void space with isolated fibrils (see, e.g., [52, 53]). The density $\rho_c$ of the craze matter and hence the extension ratio $\lambda_c$ from (2.2) change continuously from the bulk value to that of a fully developed craze with mature fibrils. This transition is modelled here by allowing $\lambda_c$ to evolve with the undeformed fibril length $\xi_0$ from the initial value $\lambda_c(\xi_0 = 0) = 1$ to a saturation value $\lambda_c^*$, representing the fully developed craze. This transition is modelled by a simple exponential ansatz depicted in Figure 2 where $\xi_{0,\max} = \lambda_c^*\, h_0$ is the relaxed fibril length at rupture and $\alpha$ defines its fraction upon which mature fibrils exist.

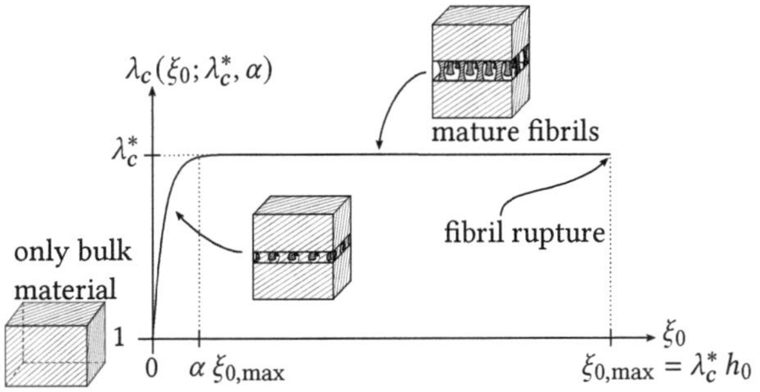


**Figure 2.2.:** Ansatz for morphology change in continuum model for $\alpha = 0.1$.

It is acknowledged that much research has been devoted to quantifying experimentally the extension ratio and that it is generally neither constant along the thickness (i.e. along a fibril) nor along the length of the craze (e.g. [49]). Nonetheless in favour of simplicity, $\lambda_c$ is considered here as constant along the craze thickness.

### 2.2.2. Micromechanics of fibril-bulk coupling

As indicated in Figure 2.1, the craze microstructure is idealised by neglecting cross-tie fibrils and considering only string-like fibrils in the direction of the maximum principal stress, which defines the coordinate system in Figure 2.1. The stress state in the fibrils is hence taken as homogeneous and uniaxial, yielding

$$\boldsymbol{\sigma}^f = \sigma^f \mathbf{e}_1 \otimes \mathbf{e}_1, \tag{2.3}$$

where $\sigma^f$ is the true (Cauchy) fibril stress. Likewise, the stress state in the bulk portion of the craze element

$$\boldsymbol{\sigma}^b = \sum_{i=1}^{3} \sigma_i^b \, \mathbf{e}_i \otimes \mathbf{e}_i, \tag{2.4}$$

is assumed to be homogeneous. Homogenisation of the layered bulk-craze structure of the element yields through the *rule of mixtures* the macroscopic Cauchy stress $\bar{\boldsymbol{\sigma}}$ in the deformed state (Figure 2.1, right) as

$$\bar{\boldsymbol{\sigma}} = \sigma_1^b \, \mathbf{e}_1 \otimes \mathbf{e}_1 + \frac{\chi}{\chi + \xi} \sum_{i=2}^{3} \sigma_i^b \, \mathbf{e}_i \otimes \mathbf{e}_i. \tag{2.5}$$

Note that changes of principal stress directions are not considered in the deformation process of interest here, as it would be the case ahead of a mode I crack as illustrated in Figure 2.1.

The overall deformation of the craze element is described by the macroscopic stretch tensor

$$\bar{\boldsymbol{\lambda}} = \sum_{i=1}^{3} \bar{\lambda}_i \, \mathbf{e}_i \otimes \mathbf{e}_i, \tag{2.6}$$

which is related to the bulk stretch tensor

$$\boldsymbol{\lambda}^b = \sum_{i=1}^{3} \lambda_i^b \, \mathbf{e}_i \otimes \mathbf{e}_i \tag{2.7}$$

and the fibril stretch in $\mathbf{e}_1$-direction $\lambda^f = \xi/\xi_0$ by

$$\bar{\lambda}_1 = \frac{\xi + \chi}{h_0} = \lambda_1^b \frac{\chi_0}{h_0} + \lambda^f \frac{\xi_0}{h_0} \tag{2.8}$$

and

$$\bar{\lambda}_2 = \lambda_2^b, \qquad \bar{\lambda}_3 = \lambda_3^b. \tag{2.9}$$

Note that, analogous to the stress as aforementioned, the deformations in the bulk phase and in the fibrils are taken as homogeneous.

Force equilibrium in $\mathbf{e}_1$-direction between bulk and fibrils reads

$$\sigma_1^b A = \sigma^f A_f \tag{2.10}$$

where $A$ and $A_f$ are the deformed unit cross-sections normal to the $\mathbf{e}_1$-plane of bulk and fibril material, respectively. The undeformed cross-sections are related to each other by the extension ratio $\lambda_c$ (cf. (2.2)) as

$$\lambda_c = \frac{A_0}{A_{f0}} \tag{2.11}$$

and in the bulk material the relation

$$A = A_0 \, \lambda_2^b \, \lambda_3^b \tag{2.12}$$

holds. For simplicity, the deformation of the fibrils is assumed to be isochoric, yielding

$$A_f = \frac{A_{f0}}{\lambda^f}. \tag{2.13}$$

Inserting (2.11) - (2.13) into (2.10) yields

$$\sigma_1^b = \frac{\sigma^f}{\lambda^f \, \lambda_c \, \lambda_2^b \, \lambda_3^b}, \tag{2.14}$$

which by virtue of (2.5) provides a coupling between the fibril stress and the macroscopic stress in $\mathbf{e}_1$-direction.

### 2.2.3. Fibril behaviour during unloading

The fibril's string-like microstructure suggests that its load-bearing capacities are essentially limited to tensile forces:

$$\sigma^f \geq 0. \tag{2.15}$$

That is, during unloading the fibrils remain traction free up to craze closure at $\bar{\lambda}_1 = 1$, and are here referred to as *loose hanging* fibrils. This however implies that the fibrils perfectly align during craze closure, resulting in a vanishing void space between the fibrils at $\bar{\lambda}_1 = 1$. The craze element's compressive response for $\bar{\lambda}_1 < 1$ is identical to the bulk material. Although there is a notion, corroborated by experimental observations [23], that compression in the craze may already occur in the macroscopic tensile regime (i.e. $\bar{\lambda}_1 > 1$) possibly due to jamming of fibrils; however, there is insufficient theoretical evidence to warrant a physically motivated refinement.

## 2.3. Constitutive equations

With the established micromechanical framework for the craze model, the constitutive description of the bulk and the fibril are discussed alongside with the criteria for craze initiation, fibril drawing and breakdown. To provide an early overview of the fibril deformation behaviour as well as and the involved quantities, the rheological model and its physical motivation is depicted in Figure 2.3. The aim in the following is to use simple constitutive models where possible, while still accounting for relevant mechanisms. To this end, the most elementary rheological representations of viscoplasticity and solid-type viscoelasticity, as depicted in Figure 2.3, are employed. Moreover, despite acknowledging that finite strain flow processes in polymers are markedly nonlinear, a simplified linear description of the rate-dependencies of fibril drawing (Subsection 2.3.2) and creep deformation (Subsection 2.3.3) through constant viscosities ($\eta_d$ and $\eta_c$ in Figure 2.3) is adopted. This allows to introduce distinct characteristic times associated with the two flow mechanisms, which strongly facilitates the parameter study focusing on their interplay during crazing under cyclic (i.e. highly transient) loading, as conducted in Section 2.4. It is noteworthy that the two viscosities, although referring to the same material, are considered to be unrelated, since the viscosity associated with fibril drawing represents chain-scission and disentanglement in the active zone. This micro-mechanism is characterised by its own time- (or rate-)dependence.

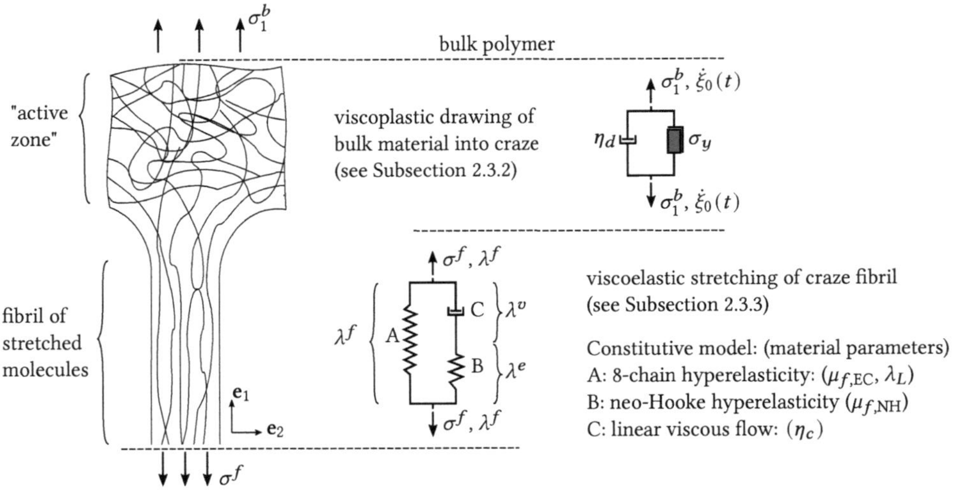


**Figure 2.3.:** Fibril model comprising viscoplastic drawing from active zone and viscoelastic fibril deformation.

### 2.3.1. Bulk deformation behaviour

For many glassy polymers (e.g. polycarbonate) the deformation behaviour prior to yielding or crazing can be well approximated by isotropic linear elasticity. Upon craze initiation, it is presumed that the bulk layer continues to deform in the linear elastic regime, given that the majority of deformation is expected within the much thinner fibrils. To still account for moderate strains of up to a few percent in that regime, the relation suggested by Anand [3]

$$\boldsymbol{\sigma}^b = 2\mu_b \left( \ln \boldsymbol{\lambda}^b + \frac{\nu_b}{1 - 2\nu_b} \operatorname{tr}\left[\ln \boldsymbol{\lambda}^b\right] \mathbf{1} \right), \tag{2.16}$$

is taken to describe the material behaviour in the bulk portion of the craze element in Figure 2.1, where $\mu_b$ and $\nu_b$ denote the bulk shear modulus and Poisson's ratio, respectively.

### 2.3.2. Craze initiation, fibril drawing and breakdown

There are various craze initiation criteria proposed in the literature, ranging from stress based [86, 87] to strain based [67] or those involving a characteristic initiation time [6, 7] (for a literature review see, for instance, [27]). Nonetheless, there is general understanding that the maximum principal stress plays a key role in the initiation process, since it determines the fibril orientation. Additionally, a certain consensus exists that hydrostatic tensile stress influences the cavitation process due to facilitation of microvoid formation. The role of the latter, however, is more ambiguous, so that for simplicity the craze initiation is associated here exclusively with a critical value of the maximum principal stress $\sigma_y$, yielding

$$\sigma_1^b = \sigma_y \tag{2.17}$$

as initiation criterion.

Based on the Argon model [5], fibril drawing is often modelled in an Eyring-type fashion, which displays a stress (or strain-rate) dependent viscosity (e.g. [26, 82, 84, 97]). In contrast, fibril drawing is described here as a viscoplastic process with a *yield* threshold $\sigma_y$ and a linear rate dependence. This approach preserves the essential feature of the drawing mechanism while simplifying the model. As initially motivated at the beginning of Section 2.3, a linear rate dependence implies a constant viscosity and consequently, a unique characteristic time. By making use of the standard notation for viscoplastic models with the Macaulay bracket $\langle ... \rangle$, fibril drawing is described by

$$\dot{\xi}_0 = \frac{h_0}{\eta_d} \left\langle \sigma_1^b - \sigma_y \right\rangle \quad \geq 0, \tag{2.18}$$

where $\eta_d$ is the drawing viscosity. Note that (2.18) ensures that fibrils originate and grow in the direction of the maximum principal stress.

Similar to craze initiation, there are a several craze breakdown criteria such as a critical deformed craze thickness [97] or a more macroscopical criterion with an accumulated plastic craze-strain [33]. Following Boyce and coworkers [82], fibril breakdown is taken here to occur upon complete consumption of the primordial thickness $h_0$, i.e. at $\chi_0 = 0$, which yields

$$\xi_{0,max} = \lambda_c^* h_0 \tag{2.19}$$

and is also depicted in Figure 2.2.

### 2.3.3. Finite strain fibril deformation behaviour

Experimental evidence by Döll and coworkers, e.g. [50], indicate that the craze thickening process under cyclic loading is a competition between drawing of new material from the active zone into the fibrils and fibril creep deformation. The latter is typically neglected in theoretical studies focussing on monotonic loading by modelling the fibril deformation behaviour as purely elastic, e.g. [26, 82, 84, 97]. Under cyclic loading, however, viscoelastic effects such as relaxation might be of relevance and hence in the following, the emphasis lies on the discussion of a proper fibril deformation model.

The viscoelastic nature of an existing craze is also corroborated by experiments conducted by Kambour and Kopp [47], who measured the cyclic stress-strain response of a polycarbonate craze in a quasi-bilinear, force-controlled loading-unloading programme. The peculiarity of the experiments is that they first created an isolated craze which was then solely strained uniaxially in such a way that no new material was drawn in. Thus, the stress-strain results depicted by the black, solid curves in Figure 2.4 provide a good notion of the fibril deformation behaviour. The craze exhibits a clear non-linear hysteresis in the first loading cycle and a tendency to a nearly elastic response upon reloading in the subsequent four cycles. After a long recovery between the $8^{\text{th}}$ and the $9^{\text{th}}$ cycle, the large initial hysteresis is to a certain degree retrieved. Very similar findings are reported for polystyrene by Hoare and Hull [38].

An initial hysteresis, the tendency to an elastic response and especially the recovery of the hysteresis, which is a strong indication of viscoelastic effects, are considered as key features

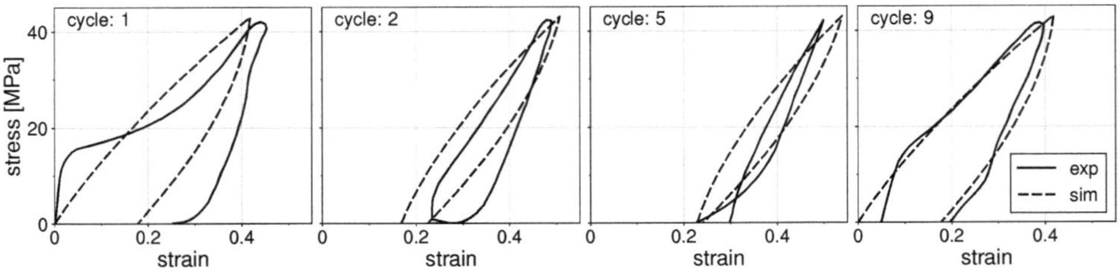


**Figure 2.4.:** Experimentally measured cyclic stress-strain response of polycarbonate craze for 5 successive loading cycles and additionally reloading after long recovery between $8^{\text{th}}$ and $9^{\text{th}}$ cycle [47] as well as the response of a least-square fitted linear viscoelastic Poynting-Thomson model.

which can be captured reasonably well by a simple linear viscoelastic Poynting-Thomson model with two elastic moduli $E_A$ and $E_B$ and the fibril viscosity $\eta_c$

$$\sigma^f = (E_A + E_B)\,\varepsilon^f - E_B \varepsilon^v, \qquad \dot{\varepsilon}^v = \frac{E_B}{\eta_c}\left(\varepsilon^f - \varepsilon^v\right), \tag{2.20}$$

as shown by the dashed curves in Figure 2.4. Although the initial hysteresis for typical glassy polymers in the first cycle, i.e. the progressive hardening upon yielding, is not replicated by the Poynting-Thomson model, in view of cyclic loading the initial behaviour appears to be less important than the subsequent response, which is approximated reasonably well considering the simplicity of the model.

As the presented crazing model is placed in a three-dimensional, finite strain setting, the one-dimensional linear viscoelastic Poynting-Thomson model needs to be adequately extended, which is addressed in the following. Recall that the fibril's stress state is uniaxial and homogeneous (cf. (2.3)) and that the fibril deformation is considered to be isochoric (cf. (2.13)). The fibril stretch tensor $\boldsymbol{\lambda}^f$ thus reads

$$\boldsymbol{\lambda}^f = \lambda^f \mathbf{e}_1 \otimes \mathbf{e}_1 + \sum_{i=2}^{3} \frac{1}{\sqrt{\lambda^f}} \mathbf{e}_i \otimes \mathbf{e}_i. \tag{2.21}$$

In accordance with the viscoelastic model shown in Figure 2.3, the fibril stretch is multiplicatively decomposed

$$\lambda^f = \lambda^e \, \lambda^v, \tag{2.22}$$

into an elastic $\lambda^e$ and viscous contribution $\lambda^v$, while the fibril stress is additively given by

$$\sigma^f = \sigma^f_{\mathrm{EC}} + \sigma^f_{\mathrm{NH}}. \tag{2.23}$$

The network stress $\sigma^f_{\mathrm{EC}}$ is modelled via the incompressible eight-chain model by Arruda and Boyce [8]

$$\sigma^f_{\mathrm{EC}} = \frac{\mu_{f,\mathrm{EC}}}{\lambda_C} \frac{\mathcal{L}^{-1}(\lambda_C/\lambda_L)}{\mathcal{L}^{-1}(1/\lambda_L)} \left(\lambda^{f\,2} - \frac{1}{\lambda^f}\right). \tag{2.24}$$

as it approximates strain hardening under uniaxial loading well [106] and has been widely used for glassy polymers at large deformations. To approximate the inverse Langevin function $\mathcal{L}^{-1}(x)$, the Padé approximation is used [20]

$$\mathcal{L}^{-1}(x) = x \frac{3 - x^2}{1 - x^2} \tag{2.25}$$

and the mean chain stretch $\lambda_C$ simplifies under uniaxial stress and isochoric deformation to

$$\lambda_C = \sqrt{\frac{\mathrm{tr}\left[\boldsymbol{\lambda}^{f\,2}\right]}{3}} = \sqrt{\frac{1}{3}\left(\lambda^{f\,2} + \frac{2}{\lambda^f}\right)}. \tag{2.26}$$

Equation (2.24) involves two material parameters, namely the initial shear modulus $\mu_{f,\text{EC}}$ and the limit stretch $\lambda_L$ corresponding to the entanglement density of the glassy polymer.

An additional nonlinear spring in series to the dashpot (Figure 2.3, element B) is modelled to enable a realistic response upon instantaneous loading. Since large elastic deformations shall be captured, the stress is given by an incompressible neo-Hookean material model

$$\sigma_{\text{NH}}^{f} = \mu_{f,\text{NH}} \left( \lambda^{e\,2} - \frac{1}{\lambda^{e}} \right), \tag{2.27}$$

which introduces a second shear modulus $\mu_{f,\text{NH}}$ as a material parameter.

As motivated at the beginning of Section 2.3, a simplified linear viscous flow model with the constant fibril creep viscosity $\eta_c$ is adopted (Figure 2.3, element C), giving rise to the flow rule

$$\dot{\lambda}^{v} = \frac{2}{3\,\eta_c} \sigma_{\text{NH}}^{f} \lambda^{v}. \tag{2.28}$$

The factor $2/3$ stems from the deviatoric part of $\sigma_{\text{NH}}^{f}$ arising from isochoric viscous flow.

### 2.3.4. Model summary and numerical aspects

The model, represented by a system of differential algebraic equations and summarised in Table 2.1, is analysed within a PYTHON environment by using contemporary computational methods for a rapid implementation and high variability for model adjustments. To this end, the automatic differentiation capabilities of the PYTORCH module [68] have been harnessed for calculating the tangent. The FSOLVE function, a component of the SCIPY module [101], has been be utilised to solve the discretized nonlinear system of equations. In our evaluation, both techniques were effectively employed and both enable fast implementations of very complex material models. This in combination with the modular structure of the craze model permits straightforward modifications. Such adjustments include, for example, incorporating Eyring-type viscous flow models.

The constitutive model has also been implemented as a user material subroutine in the finite element programme ABAQUS [1]. In that process, a modified version of the dual number automatic differentiation tool from [107] was successfully utilised, providing benefits akin to those by the PYTORCH module. Although this work does not present any finite element simulations, it is noteworthy that the tool was successfully employed on the computing cluster *bWUniCluster 2.0*, making this computational approach also a valuable tool for large boundary value problems.

**Table 2.1.:** Summary of equations governing craze model.

**Micromechanics and crazing mechanism**

| | |
|---|---|
| mass balance: | $h_0 = \chi_0 + \dfrac{\xi_0}{\lambda_c}$ |
| extension ratio: | $\lambda_c = \dfrac{\rho_b}{\rho_c}$ |
| kinematics: | $\bar{\lambda}_1 = \dfrac{\xi + \chi}{h_0} = \lambda_1^b \dfrac{\chi_0}{h_0} + \lambda^f \dfrac{\xi_0}{h_0}, \quad \bar{\lambda}_2 = \lambda_2^b, \quad \bar{\lambda}_3 = \lambda_3^b$ |
| macro stress: | $\bar{\boldsymbol{\sigma}} = \sigma_1^b\, \mathbf{e}_1 \otimes \mathbf{e}_1 + \dfrac{\chi}{\chi + \xi} \sum_{i=2}^{3} \sigma_i^b\, \mathbf{e}_i \otimes \mathbf{e}_i,$ |
| fibril stress: | $\boldsymbol{\sigma}^f = \sigma^f\, \mathbf{e}_1 \otimes \mathbf{e}_1, \quad \sigma^f \geq 0$ |
| force balance: | $\sigma_1^b = \dfrac{\sigma^f}{\lambda^f\, \lambda_c\, \lambda_2^b\, \lambda_3^b}$ |
| morphology change: | $\lambda_c(\xi_0; \lambda_c^*, \alpha)$ |
| craze initiation: | $\sigma_1^b = \sigma_y$ |
| fibril drawing: | $\dot{\xi}_0 = \frac{h_0}{\eta_d} \left\langle \sigma_1^b - \sigma_y \right\rangle \quad \geq 0$ |
| fibril breakdown: | $\xi_{0,max} = \lambda_c^*\, h_0$ |

**Fibril model**

$$\lambda^f = \lambda^e\, \lambda^v \qquad \sigma^f = \sigma^f_{\mathrm{EC}} + \sigma^f_{\mathrm{NH}} \qquad \dot{\lambda}^v = \frac{2}{3\,\eta_c} \sigma^f_{\mathrm{NH}} \lambda^v$$

$$\sigma^f_{\mathrm{EC}} = \frac{\mu_{f,\mathrm{EC}}}{\lambda_C} \frac{\mathcal{L}^{-1}(\lambda_C/\lambda_L)}{\mathcal{L}^{-1}(1/\lambda_L)} \left( \lambda^{f\,2} - \frac{1}{\lambda^f} \right) \qquad \sigma^f_{\mathrm{NH}} = \mu_{f,\mathrm{NH}} \left( \lambda^{e\,2} - \frac{1}{\lambda^e} \right)$$

$$\lambda_C = \sqrt{\frac{1}{3} \left( \lambda^{f\,2} + \frac{2}{\lambda^f} \right)} \qquad \mathcal{L}^{-1}(x) = x \frac{3 - x^2}{1 - x^2}$$

**Bulk model**

$$\boldsymbol{\sigma}^b = 2\mu_b \left( \ln \boldsymbol{\lambda}^b + \frac{\nu_b}{1 - 2\nu_b} \operatorname{tr}\left[ \ln \boldsymbol{\lambda}^b \right] \mathbf{1} \right)$$

## 2.4. Parameter study under macroscopic uniaxial strain

The aim of the subsequent parameter study is to derive a clear picture of the model's mechanisms and their mutual influence under monotonic and particularly cyclic loading. Within that scope, the crazing model is evaluated under macroscopic uniaxial strain conditions, i.e. for $\bar{\lambda}_2 = \lambda_2^b = 1$ and $\bar{\lambda}_3 = \lambda_3^b = 1$, with a prescribed, constant macroscopic stretch rate $\dot{\bar{\lambda}}_1$. In order to reduce the number of parameters and obtain results independent of a specific material, the material parameters are normalised as follows: The stiffness parameters are normalised by means of the total fibril stiffness $\mu_f := \mu_{f,\mathrm{EC}} + \mu_{f,\mathrm{NH}}$. Additionally, the characteristic times associated with fibril creep deformation $\tau_c = \eta_c/\mu_f$ and fibril drawing $\tau_d = \eta_d/\mu_f$ are cast along with the characteristic loading time $T_0 = 1/\dot{\bar{\lambda}}_1$ into the two dimensionless parameters $\tau_c/T_0$ and $\tau_d/T_0$ for creep and drawing, respectively. Increasing $\tau_c/T_0$ while maintaining $\tau_d/T_0$ enhances the fibril viscosity, whereas vice versa, elevated values of $\tau_d/T_0$ characterise a higher resistance to drawing. Increasing or decreasing both in tandem corresponds to faster or slower loading rate, respectively. The generic parameters of the crazing model are summarised in Table 2.2 and are used in the following if not indicated differently.

**Table 2.2.:** Dimensionless parameters of crazing model.

| $\mu_b/\mu_f$ | $\nu_b$ | $\mu_{f,NH}/\mu_f$ | $\sigma_y/\mu_f$ | $\lambda_c^*$ | $\alpha$ | $\lambda_L$ |
|---|---|---|---|---|---|---|
| 1 | 0.3 | 0.5 | 0.05 | 3 | 0.01 | 2 |

## 2.4.1. Monotonic loading

To investigate the effect of the drawing viscosity $\tau_d/T_0$ and the fibril creep viscosity $\tau_c/T_0$ on the craze element response, Figure 2.5(a) displays the evolution of the macro stress $\bar{\sigma}_1$, while Figure 2.5(b) depicts the progression of the relaxed fibril length $\xi_0$, both, evolving with the imposed macro stretch $\bar{\lambda}_1$. In the context of monotonic loading, the drawing viscosity and the fibril viscosity are distinguished throughout this subsection via colour coding and line style coding, respectively.

The macro stress $\bar{\sigma}_1$ (Figure 2.5(a)) undergoes a transition from a linear increase during the first few percent of strain towards an approximately constant (i.e. steady-state) plateau which depends on the different viscosities. As expected, higher stress values and accelerated rates of fibril drawing (Figure 2.5(b)) are obtained for larger viscosities, which corresponds to the response of earlier crazing models in [26, 33, 84] and observations from molecular dynamics simulations, e.g. [76]. The stress at the onset of drawing (which

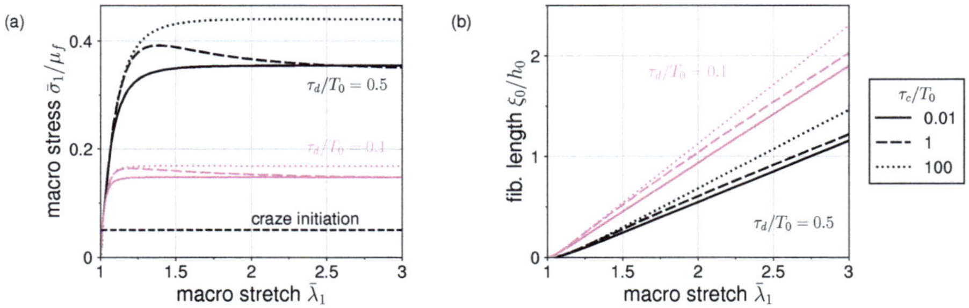


**Figure 2.5.:** Influence of fibril creep viscosity $\tau_c/T_0$ and drawing viscosity $\tau_d/T_0$ on (a) normalised macro stress $\bar{\sigma}_1$ and (b) normalised fibril length $\xi_0$.

is here identical to craze initiation, cf. (2.17)) is indicated in Figure 2.5(a). Although the initial linear response, reflecting the sole bulk deformation (cf. (2.16)), is strictly speaking confined to the pre-craze regime, Figure 2.5(a) shows that even after craze initiation the macroscopic stiffness remains approximately constant up to certain value of the macro stretch of, for instance, $\bar{\lambda}_1 \approx 1.05$ in case of $\tau_d/T_0 = 0.5$. This threshold corresponds to a critical value of the fibril length $\xi_0/h_0$ below which its contribution to the overall deformation is negligible - this might be interpreted as a *size effect*. This results directly from the craze initiation criterion (2.17), where the complex craze initiation process has been simplified to occur solely under a critical maximum principal tensile stress.

The model encompasses two deformation mechanism to comply with the imposed macro deformation, namely drawing of material into fibrillated matter of a reduced density (cf. (2.2)) and bulk/fibril deformation. For lower drawing viscosities, represented by smaller values of $\tau_d/T_0$, the craze element is inclined to comply with the imposed overall deformation by pulling in bulk material, leading to relatively small deformations in the fibrils. Consequently, the fibril creep viscosity ($\tau_c/T_0$) exerts a more noticeable impact on the macro stress when $\tau_d/T_0 = 0.5$ as compared to $\tau_d/T_0 = 0.1$. In case of $\tau_d/T_0 = 0.5$, a particular effect of an intermediate fibril creep viscosity ($\tau_c/T_0 = 1$) can also be seen in Figure 2.5(a): for high ($\tau_c/T_0 = 100$) as well as low ($\tau_c/T_0 = 0.01$) values of the creep viscosity the fibrils behave essentially elastic (with different stiffnesses) and the stress approaches its steady-state plateau value in a monotonic fashion. In contrast, the intermediate creep viscosity ($\tau_c/T_0 = 1$) gives rise to a pronounced viscoelastic fibril behaviour and the non-monotonic stress response in Figure 2.5(a) can be ascribed to the delayed fibril deformation.

Figure 2.6 provides a closer look at the deformation behaviour of the fibril, showing the evolution of fibril stress $\sigma^f$ (Figure 2.6(a)) and fibril stretch $\lambda^f$ (Figure 2.6(b)) during the overall deformation of the craze element for the same parameter values as in Figure 2.5. While the plateau of the fibril stress $\sigma^f$ is essentially determined by $\tau_d/T_0$ and is fairly independent of $\tau_c/T_0$, the fibril stretch $\lambda^f$ shows significant dependence on the fibril creep viscosity. As the fibril thins due to the continuously increasing fibril stretch for the intermediate value $\tau_c/T_0 = 1$, the fibril stress remains approximately constant or even

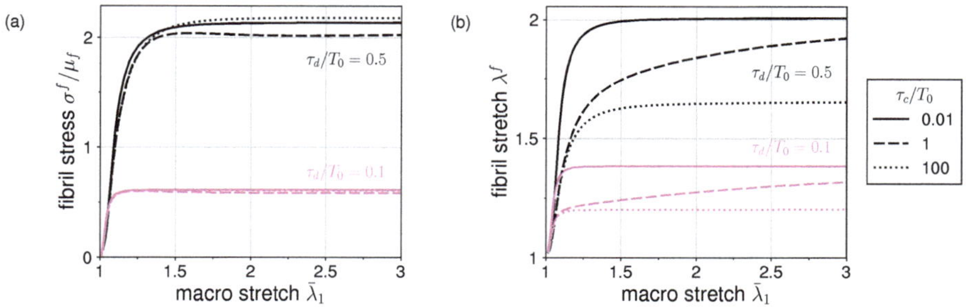


**Figure 2.6.:** Influence of fibril creep viscosity $\tau_c/T_0$ and drawing viscosity $\tau_d/T_0$ on (a) normalised fibril stress $\sigma^f$ and (b) fibril stretch $\lambda^f$.

slightly declines, resulting in a decreasing force exerted by the fibril on the bulk material and therefore a decreasing macro stress, which appears as stress-softening in Figure 2.5(a). Although this holds true for both values of the drawing viscosity, the competition between drawing and creep is more pronounced for $\tau_d/T_0 = 0.5$, as discussed above.

Comparing the macro stress plateau values in Figure 2.5(a) to the fibril stress in Figure 2.6(a) and fibril stretch in Figure 2.5(b) reveals the influence of finite strain modelling on fibril deformation. While the macro stress for the two $\tau_d/T_0$ values varies roughly by a factor of 2, the fibril stress varies by a factor of 4. This indicates, due to force equilibrium, a fibril cross-sectional area difference of a factor of 2, aligning with the fibril stretch difference (Figure 2.6(b)) given fibril incompressibility.

The morphology change from micro-voids to mature fibrils is in the model coarsely approximated by the variation of the extension ratio (cf. Figure 2.2). This variation is determined by the parameter $\alpha$ which controls the *rate* of the morphology transition. The influence of $\alpha$ on the macroscopic stress-stretch response of the craze element is depicted in Figure 2.7. A slower rate of morphology change ($\alpha = 0.1$) means a slower rate of fibril thinning and hence gives rise to a stiffer overall response. In case of the prescribed uniaxial loading of the craze element, the stress in the mature fibril regime remains essentially unaffected by the preceding morphology change. As a result, the plateau of the macro stress is nearly independent of $\alpha$ and thus, the macro stress converges to the same plateau stress regardless of the rate of morphology change, as indicated in Figure 2.7. However, the initially higher fibril stress eventually leads to the tendency for stress-softening, as also visible in Figure 2.7.

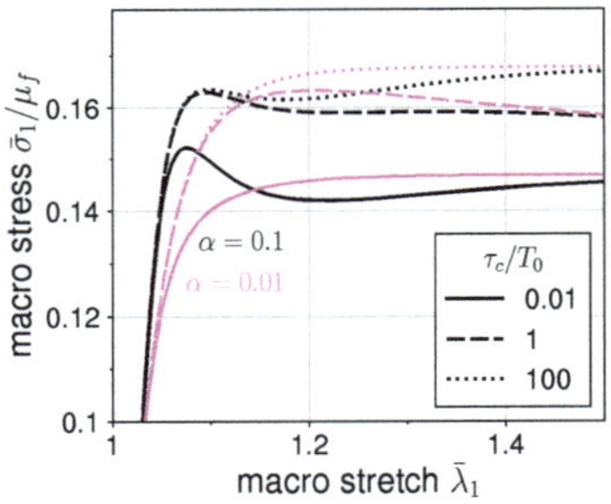


**Figure 2.7.:** Influence of rate of morphology change; $\tau_d/T_0 = 0.1$.

### 2.4.2. Cyclic loading

A cyclic loading pattern is imposed on the craze element by prescribing a macroscopic stretch history, as depicted in Figure 2.8. The stretch history is characterized by a bi-linear

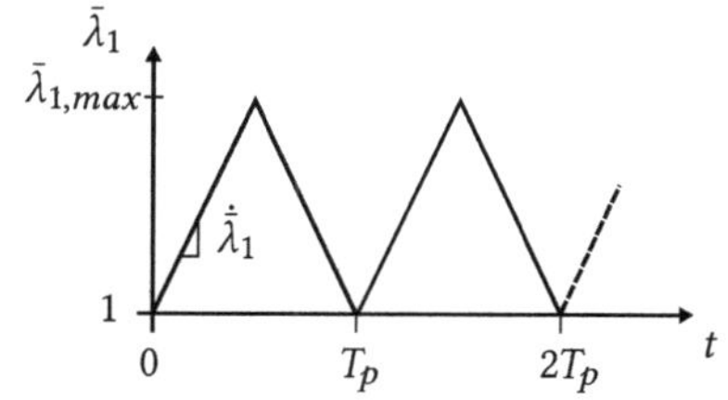


**Figure 2.8.:** Deformation controlled cyclic loading.

variation between $\bar{\lambda}_1 = 1$ (undeformed) and $\bar{\lambda}_{1,max}$ as well as the loading period $T_p$. The duration of the loading period $T_p$ is adjusted to satisfy

$$T_p = \frac{2\left(\bar{\lambda}_{1,max} - 1\right)}{\dot{\bar{\lambda}}_1} = 2\left(\bar{\lambda}_{1,max} - 1\right) T_0, \tag{2.29}$$

ensuring that the stretch rate $\dot{\bar{\lambda}}_1$ matches the one imposed under monotonic loading in Subsection 2.4.1.[1] Since the same normalisation as in Subsection 2.4.1 is adopted, results for fixed parameters are consistent across the subsections and comparisons can be drawn.

Figure 2.9 shows the temporal variation of the macroscopic stress $\bar{\sigma}_1$, the undeformed fibril length $\xi_0$, the fibril stress $\sigma^f$ and the fibril stretch $\lambda^f$ during three consecutive loading cycles for the parameters $\tau_d/T_0 = 0.5$ and $\tau_c/T_0 = 0.1$. Since the first loading stage is identical to monotonic loading (cf. Figure 2.5(a) for the qualitative evolution), the macro stress initially rises linearly and then transitions towards a plateau (Figure 2.9(a)). This is accompanied by significant fibril drawing (Figure 2.9(b)) and fibril deformation (Figure 2.9(d)). The evolution of the fibril stress (Figure 2.9(c)) follows a similar pattern as the macro stress, but as the cross-sectional area reduces due to the thinning of fibrils, the magnitude is significantly higher. Upon unloading at time $t = 0.5T_p$, both, the macro stress and the fibril stress decrease and fibril drawing ceases. This causes the undeformed fibril length $\xi_0$ to remain constant (Figure 2.9(b)). However, during this stage, the fibril stretch declines (Figure 2.9(d)) due to viscoelastic creep, which even continues once the fibrils are *loose hanging*, i.e. $\bar{\sigma}_1 = \sigma^f = 0$. Given the viscous nature of the drawing and fibril deformation mechanism, fibril drawing is reactivated upon reloading in successive cycles,

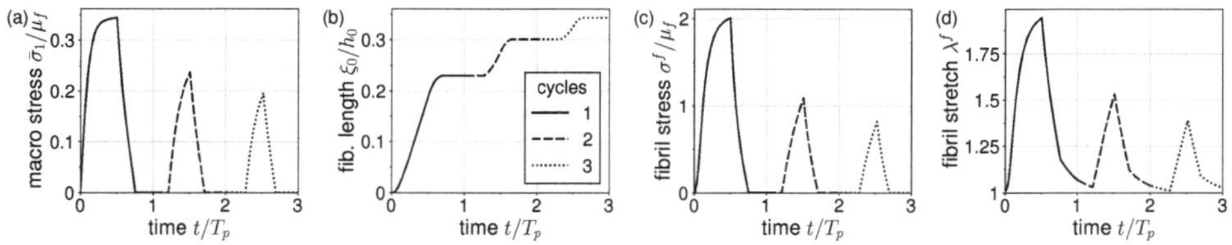


**Figure 2.9.:** Craze element response for three consecutive loading cycles; $\tau_d/T_0 = 0.5$, $\tau_c/T_0 = 0.1$.

[1] Note, a typographical error in Equation 2.29 of the original publication [56] has been corrected here.

causing further fibril growth (Figure 2.9(b)). As a consequence, on the one hand, the time intervals where fibrils are loose hanging steadily increase, as depicted in Figure 2.9(a) and Figure 2.9(c). On the other hand, the peak values of the macro stress, fibril stress, and fibril stretch drop significantly in subsequent loading cycles compared to the first cycle. Correspondingly, the amount of newly drawn fibril length $\Delta\xi_0$ per loading cycle declines with each cycle, since the macro stretch of the craze element is prescribed at a constant amplitude.

The data from Figure 2.9 are replotted in Figure 2.10 showing the macroscopic stress-stretch response of the craze element (Figure 2.10(a)) along with the fibril response (Figure 2.10(b)). The effect of creep recovery of the loose hanging fibrils can be seen in Figure 2.10 by the offset between unloading and the subsequent cycles's reloading. The corresponding hysteresis size and the the peak stress at (constant) maximum stretch decline with progressing loading cycles, which shows qualitative agreement with findings by Rabinowitz & Beardmore (cf. Figure 19 in [72]).

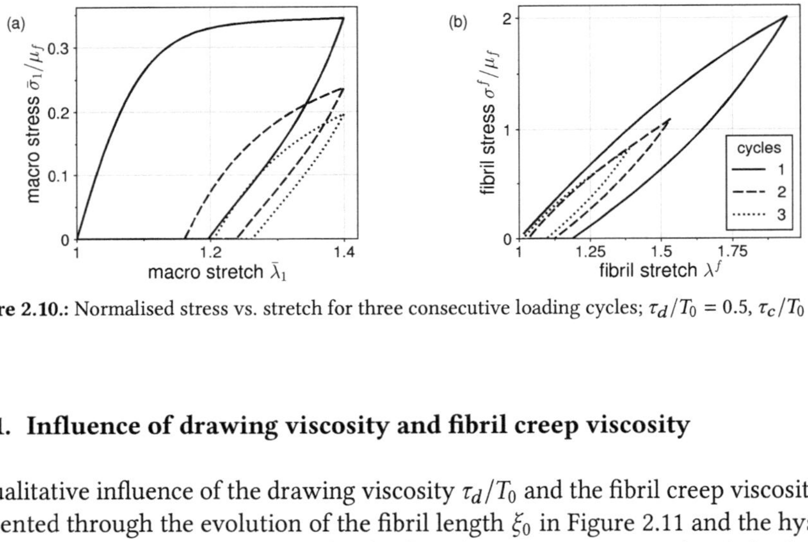


**Figure 2.10.:** Normalised stress vs. stretch for three consecutive loading cycles; $\tau_d/T_0 = 0.5$, $\tau_c/T_0 = 0.1$.

### 2.4.2.1. Influence of drawing viscosity and fibril creep viscosity

The qualitative influence of the drawing viscosity $\tau_d/T_0$ and the fibril creep viscosity $\tau_c/T_0$ is presented through the evolution of the fibril length $\xi_0$ in Figure 2.11 and the hysteretic energy dissipation $W_d$ in Figure 2.12 for the first three consecutive cycles. In both figures, the two subplots (a) and (b) depict a lower and higher drawing viscosity, respectively, whereas different fibril creep viscosities are indicated via colour coding. In case of the lower viscous drawing resistance ($\tau_d/T_0 = 0.1$) a substantial amount of fibril ($\Delta\xi_0 \approx 0.4h_0$) is drawn in the first cycle whereas only minor portions are drawn in subsequent cycles (Figure 2.11(a)). In contrast, a more balanced amount of fibril drawing per cycle is obtained for the higher drawing viscosity ($\tau_d/T_0 = 0.5$) (Figure 2.11(b)). Remarkably, the amount of fibril drawn in the second cycle relative to that drawn in the first cycle $\Delta\xi_0^{2nd}/\Delta\xi_0^{1st}$ is largest in case of the intermediate fibril creep viscosity ($\tau_c/T_0 = 0.1$), being particularly evident for the lower drawing viscosity (Figure 2.11(a)). Another interesting observation is the cross-over in the evolution of $\xi_0$ in case of $\tau_d/T_0 = 0.5$ (Figure 2.11(b)), taking place for the fibril creep viscosities $\tau_c/T_0 = 0.01$ and $\tau_c/T_0 = 0.1$. While slightly more fibril has been drawn in the first cycle for $\tau_c/T_0 = 0.01$, after the second cycle the fibril length is larger in case of $\tau_c/T_0 = 0.1$. Given that the switch does not occur in Figure 2.11(a), it implies that it results not solely from $\tau_c/T_0$, but rather from the complex interplay of fibril drawing and fibril creep in the course of craze growth under cyclic loading – at least in the scope of the present model.

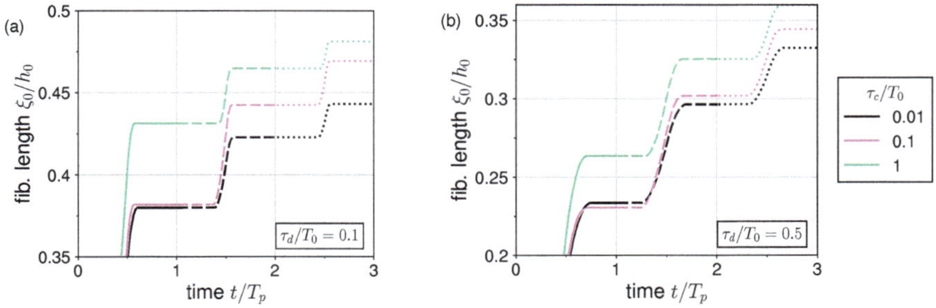


**Figure 2.11.:** Influence of drawing viscosity $\tau_d/T_0$ and fibril creep viscosity $\tau_c/T_0$ on the evolution of fibril length $\xi_0$.

The size of the stress-strain hysteresis $W_d$ after a loading cycle (Figure 2.12) corresponds to the dissipated work and can be calculated from the macroscopic cyclic stress-stretch response (cf. Figure 2.10(a)). Under uniaxial deformation, the calculation simplifies to

$$W_d = \int \bar{\sigma}_1 \frac{\dot{\bar{\lambda}}_1}{\bar{\lambda}_1} \, \mathrm{d}t. \tag{2.30}$$

For all parameter values, the dissipated work is significantly larger in the first cycle than in subsequent cycles (cf. hysteresis in Figure 2.10(a)) due to the majority of fibril drawing taking place during the first cycle under the imposed constant deformation amplitude. This results in a lower stress level which corresponds to less dissipation per cycle. The predominance of fibril drawing as initial deformation mechanism also leads to a monotonic

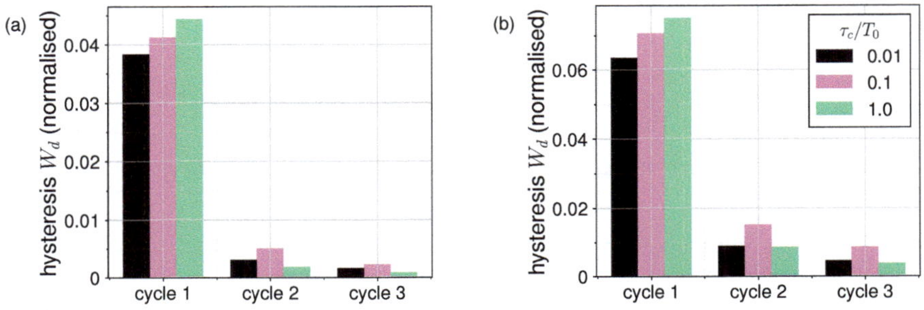


**Figure 2.12.:** Influence of drawing viscosity (a): $\tau_d/T_0 = 0.1$, (b): $\tau_d/T_0 = 0.5$ and fibril creep viscosity $\tau_c/T_0$ (colour coding) on hysteresis size over three consecutive loading cycles.

increase of $W_d$ with $\tau_d/T_0$ and $\tau_c/T_0$ in the first cycle, i.e. higher viscous resistances give rise to higher stresses and hence larger dissipation. However, particular interesting is the shift occurring in the second and third cycle where $W_d$ is largest for the intermediate fibril creep viscosity ($\tau_c/T_0 = 0.1$) for both values of $\tau_d/T_0$ (cf. Figure 2.12(a) and (b)). This change can be attributed to the influence of the then well established fibril whose viscoelastic deformation behaviour is most pronounced for the intermediate creep viscosity.

#### 2.4.2.2. Influence of morphology change $\lambda_c(\xi_0)$

To investigate the effect of the rate of morphology change (parameter $\alpha$) over multiple cycles, the imposed macro-stretch amplitude is here reduced to $\bar{\lambda}_1 = 1.1$. According to Figure 2.2, the parameter $\alpha$ corresponds to the fraction of ultimate fibril length upon which mature fibrils prevail; in terms of the normalised fibril length this state is reached at $\xi_0/h_0 (= \alpha\lambda_c^*) = 0.03$ for $\alpha = 0.01$ and at $\xi_0/h_0 = 0.3$ for $\alpha = 0.1$. Figure 2.13 illustrates this effect in terms of normalised quantities over three consecutive loading cycles. From the evolution of fibril length (Figure 2.13(c)), it can be concluded that for $\alpha = 0.01$ (as considered in previous subsections) the morphology change is completed already during the first cycle whereas for $\alpha = 0.1$ the morphology change is still ongoing even after the third cycle. While the effect on the macroscopic stress response (Figure 2.13(a)) is rather small, with slightly higher values for $\alpha = 0.1$ due to the less mature fibrils, a more pronounced influence of the parameter $\alpha$ on the evolution of the fibril length and on the fibril stress and stretch can be seen in Figure 2.13(c) and (b). The stronger fibril growth (Figure 2.13(c)) for $\alpha = 0.1$ results from the higher macro stress (essentially during the first two cycles) and hence, as the difference in peak macro stress diminishes, the amount of newly drawn fibril per cycle becomes independent of $\alpha$. Interestingly, while there is a decline in peak fibril stress and stretch across cycles for mature fibrils (Figure 2.13(b) with $\alpha = 0.01$), this trend is absent when morphological changes are still ongoing ($\alpha = 0.1$). In the latter case, peak values stay relatively consistent throughout the three cycles.

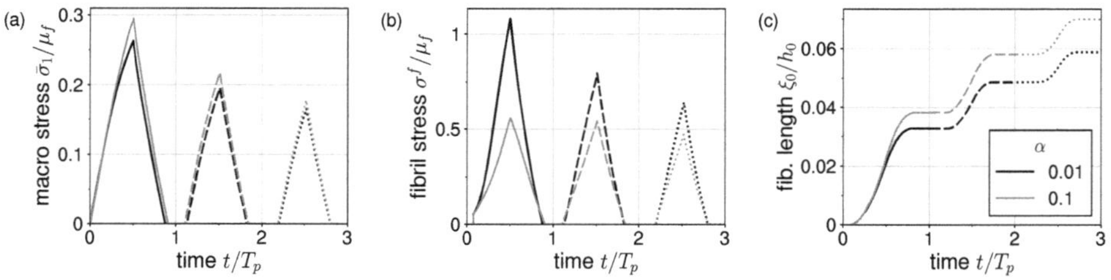


**Figure 2.13.:** Effect of rate of morphology change on craze element response over three loading cycles; $\tau_d/T_0 = 0.5$, $\tau_c/T_0 = 0.1$.

#### 2.4.2.3. Higher cyclic loading

So far, the analyses focused on the initial phases of cyclic loading. However, the response of the craze element during more advanced loading cycles also presents significant interest. To address this, the behaviour of the craze element is studied for 1000 consecutive cycles, for which the peak values (per cycle) of the macro-stress, fibril stress and fibril stretch, as well as the evolution of the fibril length are presented in Figure 2.14. A broader spectrum of fibril creep viscosities is considered, while maintaining a fixed drawing viscosity of $\tau_d/T_0 = 0.5$. Due to the constant deformation amplitude set at $\bar{\lambda}_{1,max} = 1.4$, the peak macro stress (Figure 2.14(a)) continuously decreases, converging to the drawing stress $\sigma_y = 0.05\mu_f$. This results in an arrest of fibril drawing, as depicted in Figure 2.14(b). The decline in peak fibril stress (Figure 2.14(c)) mirrors the drop in macro stress. Both these metrics become largely unaffected by the fibril creep viscosity ($\tau_c/T_0$) as steady state conditions are approached for large cycle numbers. Nevertheless, the influence of $\tau_c/T_0$ remains more pronounced in the fibril length evolution (Figure 2.14(b)) and the peak fibril stretch value (Figure 2.14(d)). In the high cycle regime, these quantities converge to two distinct values: one for $\tau_c/T_0 = 0.01$ and another for all greater creep viscosities assessed here. The observed effects, combined with the various curve intersections in Figure 2.14, highlight the complex interplay between craze fibril drawing and creep during cyclic loading. This complexity persists even under the chosen linearly viscous modelling of these mechanisms.

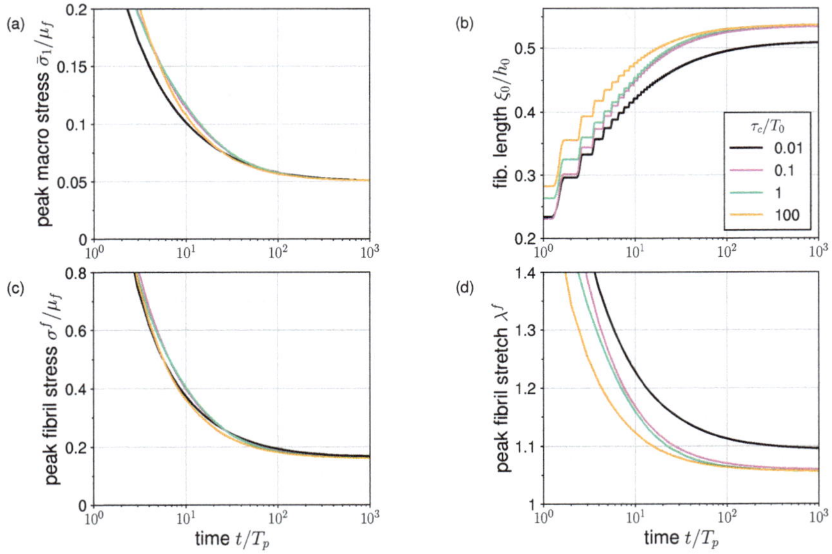


**Figure 2.14.:** Craze element response for 1000 consecutive loading cycles; $\tau_d/T_0 = 0.5$.

### 2.4.3. Fibril failure

Cyclic loading with a constant amplitude of the imposed macro-deformation has in the foregoing analyses lead to a monotonic decrease of the peak macro-stress towards the drawing stress $\sigma_y$ and hence to an arrest of fibril drawing prior to reaching its critical length for failure. In order to analyse failure of the craze element by fibril rupture and its dependence on the modelling parameters, the cyclic loading programme is now changed to one with a prescribed constant amplitude of macro-stress $\bar{\sigma}_{1,\max}$. To maintain comparability with the foregoing analyses, loading is still imposed in terms of a prescribed macro-strain rate $\dot{\bar{\lambda}}_1 = 1/T_0$, yet varying the macro-stress between zero and $\bar{\sigma}_{1,\max}$. Before looking at failure, the effect of the change in the loading programme is illustrated in Figure 2.15 where the stress-stretch response of the craze element (Figure 2.15(a)) and the evolution of the fibril length (Figure 2.15(b)) are depicted for the first four cycles. In contrast to strain-controlled loading (magenta curves, see also Figure 2.10(a)) cyclic loading with a constant macro-stress amplitude (black curves) gives rise to a *ratcheting-like* response where the hysteresis loops and the amount of newly drawn fibril length progressively increase with cycle number.

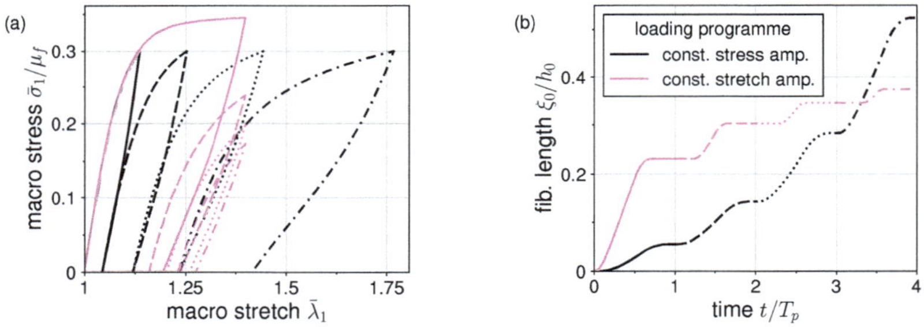


**Figure 2.15.:** Effect of stress-controlled vs. strain-controlled loading during first four loading cycles; $\tau_d/T_0 = 0.5$, $\tau_c/T_0 = 0.1$.

The overall response of the craze model up to ultimate failure is shown in terms of the peak macro-stretch per cycle and the evolution of fibril length vs. the number of loading cycles in Figure 2.16 for a macro-stress amplitude of $\bar{\sigma}_{1,\max}/\mu_f = 0.1$. It should be kept in mind that failure in our model occurs upon the complete conversion of bulk material of the primordial thickness $h_0$ into fibrils, i.e. at $\xi_0/h_0 = \lambda_c^* = 3$ according to (2.19) and Table 2.2. Again, the influence of the drawing viscosity and the fibril creep viscosity is compared via the normalised quantities $\tau_d/T_0$ and $\tau_c/T_0$ as indicated in the figures. The general trend shows an increase of the number of cycles up to failure with increasing values of both viscosities. With regard to the drawing viscosity, this effect is clearly expected owing to the higher resistance to drawing associated with higher values of $\tau_d/T_0$. However, the at first sight surprising finding that higher fibril creep viscosities give rise to larger numbers of cycles up to failure is a result of the changed loading programme. Moreover, it is worth remarking, that the fibril creep viscosity $\tau_c/T_0$ (or the fibril constitutive behaviour in

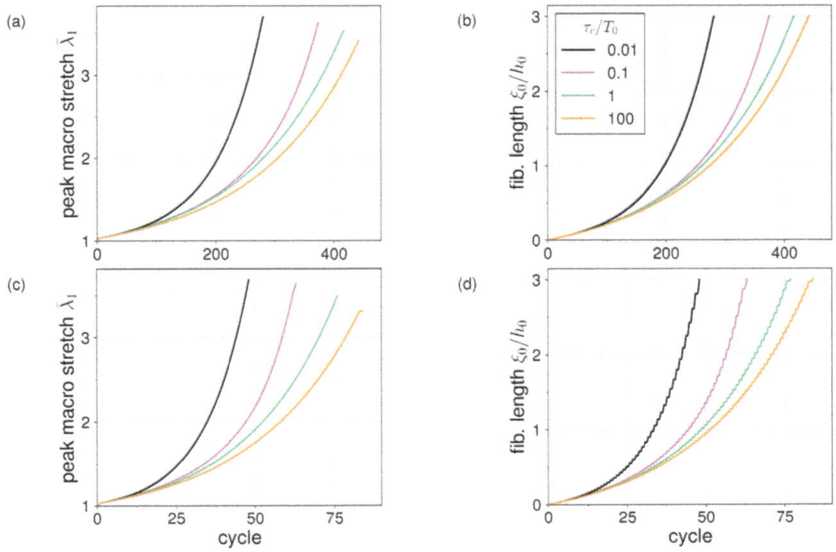


**Figure 2.16.:** Peak macro stretch and fibril length evolution $\xi_0$ for cyclic loading up to macro stress $\bar{\sigma}_1/\mu_f = 0.10$ (top: $\tau_d/T_0 = 0.5$, bottom: $\tau_d/T_0 = 0.1$).

general) has only a small influence on the overall response in the early stage of the loading programme when the fibril length is still small. This influence becomes progressively more important as the fibril length, and hence its contribution to the overall response of the craze element, increases. In other words, the fibril behaviour is only of significant influence when the fibril length is of the order of the element size (i.e. $h_0$) as can be seen in Figure 2.16. Although not explicitly illustrated here, comparing with higher values of $\bar{\sigma}_{1,\max}/\mu_f$ reveals qualitatively similar results. The notable distinction is that a larger stress amplitude essentially results in a reduced number of cycles before failure occurs.

## 2.5. Concluding remarks

The material model that has been developed here aims to represent the mechanical response of a certain *macro-element* of a craze in glassy polymers under cyclic loading normal to the overall craze plane (cf. Figure 2.1). Through rigorous continuum-micromechanical considerations the composition of the craze, encompassing both bulk material and fibrillated matter, is accommodated within a finite strain setting. The inelastic deformation behaviour of the craze element is governed by two key mechanisms: firstly, the conversion of bulk polymer into fibrillated matter of lower density by viscoplastic drawing of fibrils out of the bulk and secondly, the viscoelastic deformation of existing fibrils. Motivated by experimental findings in the literature, the second mechanism is considered to be particularly important under cyclic loading, while it is typically neglected in previous crazing models that focus on monotonic loading conditions. Indeed, the extensive parametric studies in terms of normalised material parameters within the present work have illustrated the role of fibril viscoelasticity which gives rise, for instance, to hysteretic dissipation as well as creep recovery during the unloading stages where drawing is interrupted and at which fibrils are considered *loose-hanging* and stress-free. The model has been kept as simple as possible to analyse the effect of cyclic loading on the evolution of the craze microstructure and its overall response up to failure. To this end, the inelastic mechanisms have been modelled via simplified linear rate dependencies, i.e. by constant viscosities, though it is well known that real polymer behaviour is more complex. Nonetheless, more realistic, e.g. Eyring-type, flow models can easily be incorporated with the discussed computational techniques and the modular structure of the material model. Further important simplifications in our model concern the negligence of cross-tie fibrils which might induce a non-uniaxial stress state in the fibrillated matter. Also, the criterion for craze initiation here involves only the bulk normal stress and ignores the influence of hydrostatic stress. Nevertheless, according to the parameter studies in the present work, the model predicts craze failure after a certain number of loading cycles which depends on material parameters and loading characteristics in a meaningful manner. The study shows that while the fibrils are still small compared to the primordial thickness $h_0$ of the craze element, the contribution of the viscoelastic fibril deformation to the macroscopic response is overshadowed by the drawing viscosity $\tau_d/T_0$. Yet, with increasing fibril length in the later stages of a loading programme up to failure, the influence of the fibril constitutive behaviour and hence the competition between drawing and creep becomes progressively more pronounced. This effect is considered to be independent of the constitutive description of the fibril model, i.e. the chosen rheological representation.

An apparent shortcoming of this study is the lack of a direct comparison of the presented results to experimental findings. While this has to be seen in the light of the limited availability of experimental studies characterising the response of a craze under cyclic loading, it appears necessary to discuss the qualitative predictive capabilities of the proposed model. Firstly, the findings by Kambour & Kopp [47] and Hoare & Hull [38], which focus on the behaviour of an existing craze under cyclic stress, motivate to model the fibrillated craze matter as viscoelastic, as shown in Figure 2.4. Moreover, the decrease of peak stress in the craze response under cyclic loading with a constant overall strain amplitude in Figure 2.10(a) corresponds qualitatively to experimental results by Rabinowitz & Beardmore [72] (their Fig. 19) for the same type of loading. Apart from these rare experimental data, nowadays molecular dynamics (MD) simulations provide an additional insight into the micromechanics of crazing, yet so far only under monotonic loading conditions, e.g. [32, 76, 100]. Comparison of their results with our Figure 2.5(a) shows a qualitative agreement in the constant stress plateau after craze initiation, but the pronounced stress drop right at craze initiation featured by the MD simulations is not captured by our model. This is due to the fact that the stress redistribution (essentially from hydrostatic to uniaxial) during void formation and fibrillation in the neat bulk polymer is not accounted for in the present model. Unclear, however, is whether this significant stress drop actually occurs in the advance of an existing craze, which is understood to proceed by the so-called *meniscus-*

*instability* ([7]). Concluding, although the proposed model integrates key mechanisms based on experimental findings, a comprehensive characterisation of the craze response under cyclic loading, particularly across a wide range of loading conditions, remains a challenge.

The craze model has been formulated in a way that it can readily be used for the simulation of craze and crack propagation in glassy polymers under cyclic loading where it can be interpreted as a *traction-separation law* of a cohesive zone model. In such a crack propagation study, the finite strain setting of the model also allows the integration of bulk polymer models describing shear yielding, enabling the thorough investigation of interesting phenomena arising from the interaction between crazing and shear yielding during fatigue crack growth in glassy polymers. This study will be addressed in future work.

# 3. Molecular simulations of crazes in glassy polymers under cyclic loading

**This chapter reproduces:**[*]
Laschuetza T, Ge T, Seelig T, and Rottler J. "Molecular Simulations of Crazes in Glassy Polymers under Cyclic Loading". In: *Macromolecules.* 57(23): 10894–10902, 2024. DOI: 10.1021/acs.macromol.4c01445

**Abstract:**
We study with molecular dynamics simulations of a generic bead-spring model the cyclic crazing behaviour of glassy polymers. The aim is to elucidate the mechanical response of sole fibrillated craze matter as well as its interaction with bulk material. The macroscopic stress response exhibits a hysteresis, which is quasi stationary after the first cycle and largely independent of deformation rate and temperature. It results from a complex interplay between constraints imposed by the entanglement network, pore space and pore space closure. Once the craze fibrils are oriented, stretching of the covalent backbone bonds leads to a rapid stress increase. In the initial stages of unloading, a loss in entanglement contact yields a quick stress relaxation in the backbone. During unloading, the craze fibrils undergo a rigid body (i.e. stress-free) folding motion due to the surrounding pore space, so that the structural behaviour of craze fibrils during unloading is most accurately described as string-like. The reloading response depends significantly on the degree of pore space closure and the enforced intermolecular interaction during unloading. It ranges from a linear stress increase to a recavitation with a redrawing response. Compared to the bulk stiffness, the craze stiffness is two orders of magnitude lower and as a result, the macro response of coexisting craze and bulk matter is governed by the sole fibrillated craze matter.

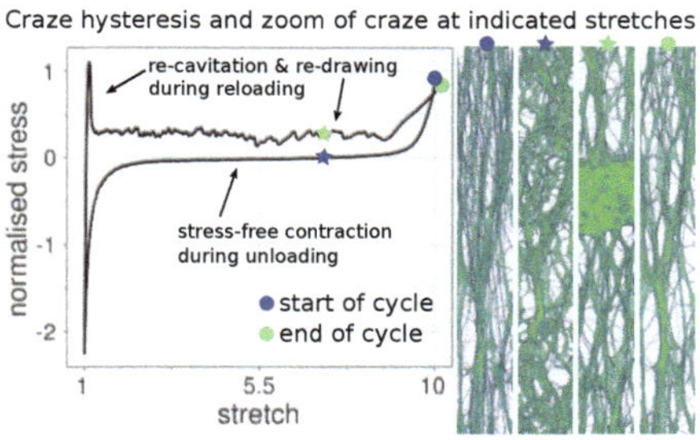


**Keywords:** Craze, Cyclic loading, Molecular dynamics simulation

## 3.1. Introduction

Crazing refers to the dilatant process of localized formation and growth of crack-like defects in glassy thermoplastic polymers. A craze consists of fibrillated matter with an interconnected void space. The several tens of nanometre thick fibrils can grow up to a few micrometres in length by drawing in surrounding bulk material from the so-called active zone [53]. This process is of practical importance since, unlike cracks, craze fibrils enable a considerable load transfer between the craze surfaces and substantially enhance the fracture toughness. Therefore, much research has been devoted to understanding the governing mechanisms, cf. reviews in [34, 46, 48, 49], which includes theoretical studies on a continuum scale, e.g. [12, 26, 33, 35, 80, 84, 97], as well as on a molecular scale, e.g. [11, 22, 31, 32, 59, 60, 65, 74–76, 98, 100, 102]. However, the focus of those theoretical studies is limited to monotonic loading conditions. In contrast, extensive experimental research for cyclic (especially fatigue) loading exists, cf. reviews in [23, 78, 94], providing insight into very interesting fracture processes for several glassy polymers and at several loading amplitudes: For instance, at low loading amplitudes, Skibo et al. [83] observed discontinuous crack growth, which was attributed to craze thickening resulting from a competition of fibril drawing and fibril creep deformation by Könczöl et al. [50]. An increase in loading amplitude may give rise to the formation of shear bands and their interaction with crazes, leading to discontinuous epsilon-shaped fatigue cracks as studied by Takemori [94]. Yet, the involved length scales pose difficulties to explore the driving mechanisms by solely relying on experiments and without theoretical analyses based on a physically motivated craze model. For this purpose, a continuum micromechanical model was recently developed by some of us [56] focusing on the cyclic response of craze matter. The model describes the structural response of craze fibrils as string-like and accounts for viscoplastic fibril drawing and viscoelastic fibril deformation, where the latter is motivated by experimental observations [38, 47]. It is designed to be employed as a traction separation law in a mode I boundary value problem to investigate cyclic craze and crack growth. The continuum scale also accounts for inelastic shear yielding in the surrounding bulk, which enables the analysis of its competition with crazing. However, the craze model suffers from two shortcomings arising from a general knowledge gap of the cyclic craze response: On the one hand, the structural behaviour of craze fibrils during unloading is uncertain. The correct structural behaviour is important since it impacts the pictures of the damage mechanism during cyclic (fatigue) loading. On the other hand, there is a lack in detailed mechanical knowledge regarding the response of craze deformation for a broad range of loading conditions.

In this paper, the response of a craze under cyclic loading is studied based on molecular dynamics simulations of a coarse-grained bead-spring model. In the past, this approach has yielded valuable insight into many molecular scale mechanisms of crazing, in particular the ones concerning the entanglement network. General studies of craze nucleation [76], craze fibril drawing [11, 59, 75, 100], and craze fracture [74] in linear flexible polymers were extended to nanocomposites [98], void nucleation [60], semiflexible and stiff polymers [22, 65], and assessed the role of the entanglement network in setting the craze extension ratio

[32, 65]. Recent simulations have gone beyond the generic coarse-grained bead-spring model and used a scale-bridging approach to simulate craze formation in polystyrene [102]. Contrary to the energy landscape and stress level, the central role of the entanglement network in controlling the structural features of craze fibrils is independent of the degree of coarse graining.

The aim in this work is twofold: First, to the knowledge of the authors, molecular dynamics simulations have not yet been employed to study the cyclic response of fibrillated craze matter. The access to molecular scale details allows the simulations to establish the microscopic picture, which is often assumed in micromechanical models but is difficult to evaluate experimentally. Therefore, this study aims to elucidate the mechanical behaviour under cyclic loading and the underlying mechanisms leading to the macroscopic response. To this end, different loading conditions are analysed for sole fibrillated craze matter as well as for systems where bulk and craze matter coexist. The latter is of interest, since crazing is a transient process which gains particular relevance during cyclic loading where, for instance, fibril drawing takes place over multiple cycles. A further focus concerns the structural behaviour of craze fibrils during unloading, and whether it can be characterised as string-like. Second, findings from this study aim to be transferred as a bottom-up approach to larger scale models to create, for instance, a molecular dynamics informed continuum model.

The presented paper is outlined as follows: In Section 3.2, the computational methods featuring the model, the system setup and the crazing simulation are detailed. The results are presented in Section 3.3 comprising the cyclic response of sole fibrillated craze matter (Subsection 3.3.1) as well as the cyclic response of coexisting craze and bulk material for different craze-bulk compositions (Subsection 3.3.2). Concluding, the key findings are summarised in Section 3.4.

## 3.2. Model and methods

The generic bead-spring model [54] is employed to study the cyclic craze response. The model setup is consistent with previous studies on crazing, e.g. [11, 31, 32, 74–76], and is briefly summarised: Each polymer is modelled as a linear chain of $N$ spherical monomers with a mass $m$. Non-bonded monomers separated by the distance $r$ interact via the truncated and shifted 6-12 Lennard-Jones (LJ) potential

$$U_{\mathrm{LJ}}(r) = 4u_0 \left[ \left(\frac{a}{r}\right)^{12} - \left(\frac{a}{r}\right)^{6} - \left(\frac{a}{r_c}\right)^{12} + \left(\frac{a}{r_c}\right)^{6} \right] \qquad \text{for} \quad r \leq r_c \tag{3.1}$$

with a cutoff radius $r_c = 1.5a$. The results are expressed in terms of Lennard-Jones units, featuring the characteristic length $a$, energy scale $u_0$ and a characteristic time $\tau = a\sqrt{m/u_0}$.

The bonded monomers interact via an attractive finitely extensible nonlinear elastic (FENE) potential with a purely repulsive second Lennard Jones term ($r_c = 2^{1/6}a$)

$$U_{\text{FENE}}(r) = -\frac{1}{2}kR_0^2 \ln\left[1 - \left(\frac{r}{R_0}\right)^2\right] + 4u_0\left[\left(\frac{a}{r}\right)^{12} - \left(\frac{a}{r}\right)^6\right] \tag{3.2}$$

with the parameters $R_0 = 1.5a$ and $k = 30u_0/a^2$, allowing entanglements to form [54]. Chain scission is not considered, since the focus lies on cyclic loading resulting in stress states lower than necessary for chain scission to occur [76].

The system comprises $M = 1800$ chains where each chain consists of $N = 500$ beads with an average entanglement length of $N_e \approx 80$ beads. Simulations are performed with the LAMMPS molecular dynamics code [69, 95], where periodic boundary conditions are employed along all three directions.

The melt state was constructed by generating random-walk coils with a subsequent equilibration with $r_c = 1.12a$ at temperature $T = 1u_0/k_b$. To facilitate equilibration, a double bridging algorithm [10] is utilized. The temperature was controlled via a Nosé-Hoover thermostat with a damping rate $1\tau^{-1}$. While retaining the volume, the equilibrated system is then quenched with $r_c = 1.5a$ to $T = 0.49u_0/k_b$, leading to zero pressure. Further quenching to the final target temperature $T = 0.1\,u_0/k_b$ takes place at zero pressure by using a Nosé-Hoover barostat with a damping rate $0.1\tau^{-1}$. The quenched system has initial box dimensions $L_{x0} = 94.3a$ and $L_{y0} = 55.8a$ in lateral direction and $L_{z0} = 167.3a$ in axial direction. All directions sufficiently exceed the end-to-end distance of the chains, which reduces finite size effects [41, 76]. The larger box length along the z-direction minimises further the effects of the finite box size on the coexistence of the craze fibrils and uncrazed regions. $L_{z0}$ being larger than the active zone size allows better separation of the craze fibrils from the active zone at the same axial stretch $\lambda_z = L_z/L_{z0}$, which is measured with respect to the isotropic glass. The box sizes in the periodic x- and y-directions are reduced compared to $L_{z0}$, so the simulations with the same number of degrees of freedom can focus on the more interesting z-direction.

Following previous studies, e.g. [32, 74, 76, 102], crazing is induced by subjecting the system to an affine uniaxial deformation along the $z$-axis, i.e. lateral stretches $\lambda_x = \lambda_y = 1$, while $\lambda_z$ is prescribed via a constant material strain rate $\dot{\lambda}_z = \dot{u}/L_{z0} = 2.6 \cdot 10^{-4}\tau^{-1}$. To simulate cyclic loading conditions, a bi-linear loading/unloading program is imposed by simply switching the velocity direction upon reaching the maximum stretch $\lambda_{z,max}$, leading to the loading period $T_p$. Throughout the loading program, a Nosé-Hoover thermostat with a damping rate $1\tau^{-1}$ is applied to the lateral velocities to maintain a constant temperature.

To study the cyclic response of sole craze material in Subsection 3.3.1, the isotropic glass is deformed to $\lambda_{z,max} = 10$. This ensures complete conversion of bulk material into fibrillated craze matter, which takes place when the deformation reaches approximately the extension ratio $\lambda_c$ (i.e. $\lambda_z \approx \lambda_c = \rho_b/\rho_c$), describing the ratio of bulk $\rho_b$ and craze density $\rho_c$. It also avoids the deformation state where chain breakage occurs (cf. [76]). In

Subsection 3.3.2, the bulk-craze interaction for different bulk-craze compositions is studied by varying $\lambda_{z,max}$. To analyse a broad response range, the unloading magnitude $\lambda_{z,u}$, i.e. the stretch to which the system is unloading, is varied in both studies. In all considered scenarios, the cycle count and the cycle time $\tau_c = \tau/T_p$ commences upon reaching $\lambda_{z,max}$ for the first time after loading from the isotropic glass.

## 3.3. Results

### 3.3.1. Cyclic response of fibrillated craze matter

We first aim to elucidate the mechanical response of sole craze matter by focusing on the macroscopic behaviour throughout the loading cycles and thereafter analysing the underlying mechanisms. Figure 3.1 presents the macroscopic response of fibrillated craze matter under cyclic loading for three simulations to $\lambda_{z,u} = 1$ (dash-dotted line), $\lambda_{z,u} = 2$ (dashed line) and $\lambda_{z,u} = 5$ (solid line). The deformation controlled cyclic loading program is shown in Figure 3.1(a), and the corresponding macroscopic axial stress $\sigma_z$ as function of the macro stretch $\lambda_z$ is depicted in Figure 3.1(b). The grey dotted line represents the initial response of the isotropic glass during craze formation and growth, featuring an elastic stress increase, the subsequent stress drop during cavitation, as well as a constant stress plateau as craze fibrils are drawn from the bulk. This monotonic crazing process has already been thoroughly studied, e.g. [11, 32, 74, 76]. Here we are interested in the mechanical response of the craze (fibrils) under cyclic loading (black lines). To provide a better visualisation, Figure 3.1(c) shows a zoom of the stress-stretch response (red box in Figure 3.1(b)), and the arrows in Figure 3.1(c) show the loading direction. The colour coding and the roman numerals indicate different craze stages.

Before discussing the different stages of the craze deformation cycle, several key features are briefly summarised. During unloading from $\lambda_{z,max} = 10$ (indicated by the vertical arrow in Figure 3.1(c), the craze first exhibits a rapid stress decline which transitions into a long, quasi-stress free contraction of the craze (orange line). Note that this unloading behaviour is identical for all three simulations, i.e. the solid, dashed, and dash-dotted curves collapse onto each other. When the stretch is reduced below $\lambda_{z,u} < 2.5$, the quasi stress-free contraction becomes compressive, and the magnitude of $\sigma_z$ significantly rises as $\lambda_z$ approaches $\lambda_z = 1$ (dash-dotted line), i.e. the macroscopically undeformed state of the initial isotropic glass. In that particular case, $\sigma_z$ increases rapidly during reloading and the craze exhibits what appears to be recavitation and reyielding, while both stress magnitudes remain below the cavitation and drawing stresses during the initial craze creation (grey dotted line). By contrast, for the case of $\lambda_{z,u} = 2$ (dashed line) and $\lambda_{z,u} = 5$ (solid line), the craze exhibits a plateau stress and a linear increase, respectively, during reloading (cf. blue dashed and solid line in Figure 3.1(c)). Around $\lambda_z \approx \lambda_c$, which is indicated by the vertical line in Figure 3.1(c), the stress rapidly grows and the response becomes nearly independent of $\lambda_{z,u}$. Moreover, it is notable that all load program variations lead

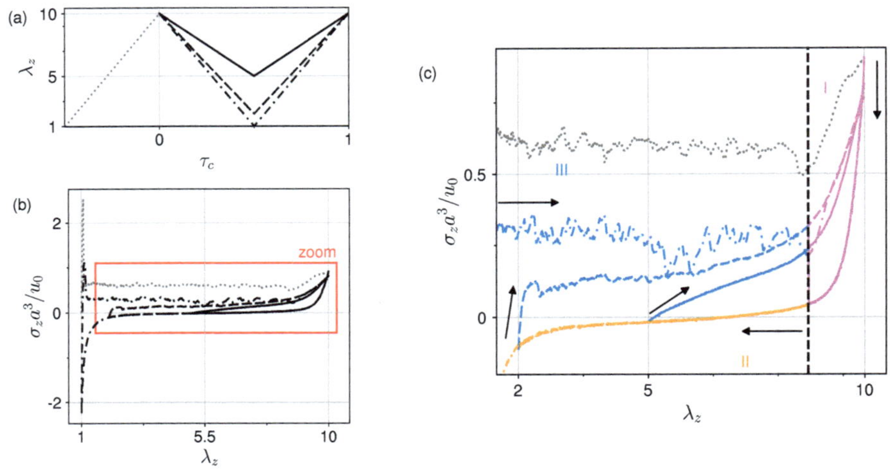


**Figure 3.1.:** Mechanical response during initial craze formation (grey dotted line) and subsequent cyclic loading (black lines) to various unloading stretches of $\lambda_{z,u} = 1$, $\lambda_{z,u} = 2$ and $\lambda_{z,u} = 5$, represented by black dash-dotted, dashed, and solid lines, respectively. (a) Schematic loading program, (b) axial stress $\sigma_z$ as function of axial stretch $\lambda_z$ with red box indicating zoom shown in (c). The arrows and colour coding in panel (c) show loading direction and different craze stages, respectively. The latter are described in the text.

to a hysteresis, which appears to be neither a thermally activated nor a rate-dependent process and relatively stationary after the first cycle (cf. Figure 3.B.1 and Figure 3.B.2 in the supporting information).

To analyse the driving mechanisms leading to the hysteresis, it is beneficial to distinguish between different craze stages during the deformation cycle. As aforementioned, the stages are represented by the colour-coding and the roman numerals in Figure 3.1(c) and can be summarised as follows:

- Stage I (magenta): Stretching and unloading of highly oriented craze fibrils. This is labelled here as first stage, since the response is essentially independent of the previous loading history.
- Stage II (orange): Quasi stress-free contraction of the craze with a transition to compression around $\lambda_z = 2.5$.
- Stage III (blue): A reloading response which depends on $\lambda_{z,u}$ and exhibits either a reyielding ($\lambda_{z,u} \leq 2$) or a linear ($\lambda_{z,u} \geq 5$) behaviour.

To further enhance the understanding, animations (created with OVITO [89]) of the three simulations are provided in the supporting information. The following analysis focuses on the 1st cycle as defined in Figure 3.1(a), if not indicated otherwise.

#### 3.3.1.1. Mechanics during stage I

The mechanics in stage I are discussed first while focusing on the mechanisms leading to the rapid stress increase during reloading, and the difference between loading and unloading that results in the hysteresis. The former is investigated through a decomposition of $\sigma_z$ into its intramolecular $\sigma_{z,b}$ and intermolecular $\sigma_{z,p}$ contributions, as shown in Figure 3.2 for $\lambda_{z,u} = 5$. Since the kinetic contribution is small in the glass, it is omitted here. Stress decompositions for the cases $\lambda_{z,u} = 2$ and $\lambda_{z,u} = 1$ are shown in Figure 3.B.3 and exhibit qualitatively similar trends. Focusing only on stage I (i.e. right hand-side of vertical dashed line), it can be concluded that the rapid stress change is governed by $\sigma_{z,b}$ and hence, by the deformation of the polymer backbone.

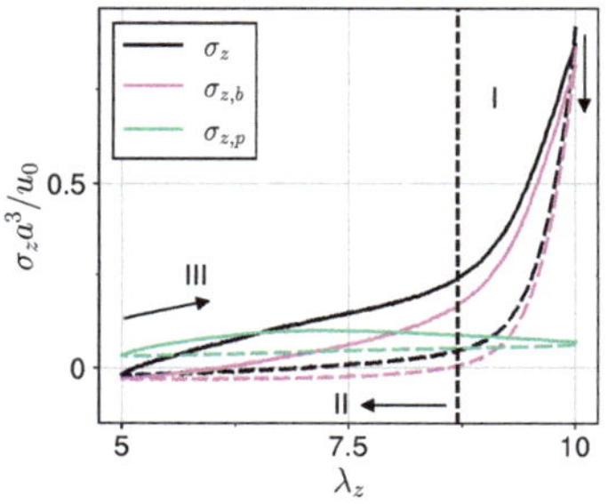


**Figure 3.2.:** Axial stress $\sigma_z$ decomposition into intermolecular pair $\sigma_{z,p}$ and intramolecular bond $\sigma_{z,b}$ components during unloading (dashed lines) and reloading (solid lines) for $\lambda_{z,u} = 5$. Arrows show load direction and roman numerals indicate craze stages.

Since Figure 3.2 suggests that chain (re)orientation is important, we further analyze the changes in chain configuration by computing orientation vectors $\mathbf{R}(N_i, \bar{\lambda}_1)$ between $N_i$ beads along the chains for a given stretch $\lambda_z$. The average orientation of these vectors is conveniently described by

$$S_z = \langle P_2(\cos \alpha_z(N_i, \lambda_z)) \rangle , \tag{3.3}$$

where $P_2(x) = (3x^2 - 1)/2$ is the second Legendre polynomial, $\alpha_z$ the angle formed by $\mathbf{R}(N_i, \bar{\lambda}_1)$ with the deformation axis and $\langle ... \rangle$ describes the ensemble average. Figure 3.3 shows a parametric plot of the stress as function of $S_z$ for five values of $N_i$, where $N_i = 80$ coincides with the entanglement length $N_e$.

The first notable observation is that $S_z$ is not an adequate order parameter at any length scale $N_i$ to characterise the hysteresis, since the hysteresis loops do not collapse. Second, given that $S_z(N_i = 80) \approx S_z(N_i = 170)$, the orientation seems to be enforced by the entanglement network. Third, before unloading from $\lambda_{z,max}$, the chains are highly oriented, which is essentially retained during the initial stages of unloading (dashed lines) in which the stress rapidly decays. Lastly, the rapid stress rise during reloading (solid lines) in regime I is accompanied by a minor change in orientation.

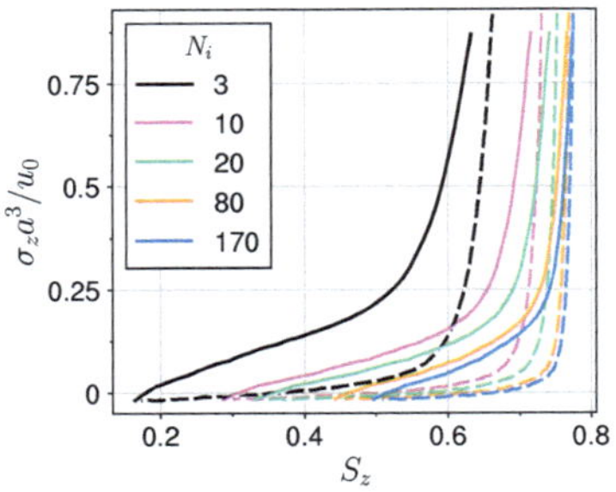


**Figure 3.3.:** Axial stress $\sigma_z$ vs. bond orientation $S_z$ for several length scales encompassing $N_i$ monomers. Data is shown during unloading (dashed lines) and reloading (solid lines) for $\lambda_{z,u} = 5$.

The stretching and relaxing of the backbone bonds is further corroborated by Figure 3.4, which shows the average resultant tensile bond force $\langle F_b \rangle = \langle \, \mathrm{d}U_{\mathrm{FENE}}(r) / \, \mathrm{d}r \rangle$. It closely follows the bond stress $\sigma_{z,b}$ derived from the virial stress, and paints the picture that constraints such as the entanglement network leads to a rise in $\sigma_z$ once chains become highly oriented. Furthermore, upon unloading, the loss in entanglement contacts yields a swift stress relaxation in the backbone and thus causes the difference in loading and unloading.

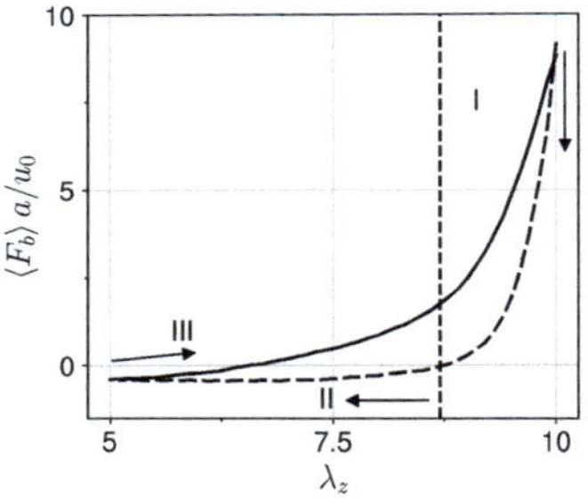


**Figure 3.4.:** Average backbone bond force $\langle F_b \rangle$ vs. stretch $\lambda_z$ during unloading (dashed lines) and reloading (solid lines) for $\lambda_{z,u} = 5$. Arrows show load direction and roman numerals indicate craze stages.

In conclusion, stage I is governed by the stretching and relaxing of the highly oriented bonds due the entanglement network. As the bonds form craze fibrils, it equivalently can be stated that this stage is governed by the deformation of the highly oriented craze fibrils. Given the necessary pre-orientation in order for the deformation to take place, this stage is essentially independent of the prior unloading history (i.e. of $\lambda_{z,u}$).

### 3.3.1.2. Mechanics during stage II

Stage II focuses on the mechanisms leading to the quasi stress-free contraction of the craze during unloading as well as the transition to compression for $\lambda_z \leq 2.5$. The data presented so far highlights two things in this regime: Firstly, the largest reduction in chain orientation occurs during this stage (cf. Figure 3.3) and secondly, $\sigma_{z,b} \approx \langle F_b \rangle \approx 0$ (cf. Figure 3.2 and Figure 3.4). The latter suggests that the structural behaviour of craze fibrils is string-like rather than beam-like as the craze (pore space & fibrils) solely exhibits a minor stiffness during contraction. This also implies a negligible bending stiffness of the craze fibrils, which validates a recent assumption on the structural behaviour of craze fibrils used in a continuum model [56].

Besides the chain level observables, a further helpful quantity to characterise the response is the particle motion orthogonal (i.e. in $xy$-plane) to the deformation axis ($z$-axis). Since $\lambda_x = \lambda_y = 1$, any such motion is nonaffine. The average lateral particle displacement $\langle u_{xy} \rangle$ is given by

$$\langle u_{xy} \rangle = \langle |\mathbf{r}_{xy}(\tau_c) - \mathbf{r}_{xy}(\tau_c = 0)| \rangle , \tag{3.4}$$

where $\mathbf{r}_{xy}$ is the position vector in the $xy$-plane and $\langle u_{xy} \rangle$ is the displacement with respect to $\lambda_{z,max}$ at the beginning of the cycle at $\tau_c = 0$. Figure 3.5 displays the evolution of $\langle u_{xy} \rangle$ (black line) as well as $\sigma_z$ (magenta line) throughout the loading cycle. During unloading ($0 < \tau_c \leq 0.5$), $\langle u_{xy} \rangle$ increases sharply. That is, the lateral particle movement increases while the bond orientation decreases, which can also be seen in the supplementary animation as a *folding mechanism* of the craze fibrils. The pore space plays a key role as it enables the motion of fibril folding to occur essentially as a rigid body motion, i.e. stress-free. It is therefore concluded, that as long as pore space exists, the structural behaviour of craze fibrils is thus most accurately described as string-like.

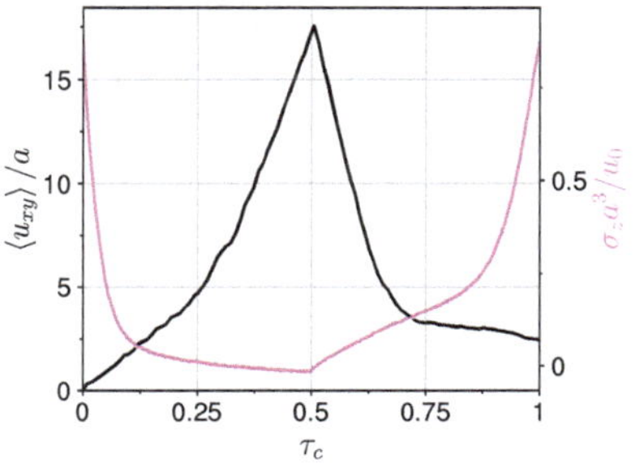


**Figure 3.5.:** Lateral particle displacement $\langle u_{xy} \rangle$ (left y-axis) and $\sigma_z$ (right y-axis) vs. cycle time $\tau_c$ for $\lambda_{z,u} = 5$.

The compressive stress increase observed for $\lambda_z \leq 2.5$ (cf. Figure 3.1(b)) arises from an increase in $\sigma_{z,p}$ (cf. stress decomposition of $\lambda_{z,u} = 1$ and $\lambda_{z,u} = 2$ in Figure 3.B.3). Although the magnitude is quite large and most likely strongly influenced by the high deformation speed in MD simulations, this behaviour is not unexpected due to the macroscopically observed dilatant deformation during craze formation and growth, leading to compressive stress ($\sigma_z < 0$) during unloading while the macroscopic deformation is still tensile ($\lambda_z > 1$). Compression at the crack tip, i.e. where the craze fibrils are the largest, was also computed using the experimentally measured craze contour as input in a finite element simulation [16, 50].

#### 3.3.1.3. Mechanics during stage III

Stage III features the reloading response, which significantly depends on $\lambda_{z,u}$. The supplementary animations show that in the case of $\lambda_{z,u} = 5$, the craze morphology is largely retained throughout the cycles. By contrast, $\lambda_{z,u} = 2$ yields a section-wise unfolding of the craze and $\lambda_{z,u} = 1$ even a redrawing. The previous data implies that pore space closure and the exerted compression highly influence the reloading behaviour. To further study the underlying mechanism that contributes especially to the difference between $\lambda_{z,u} = 2$ and $\lambda_{z,u} = 1$, the simulation box length $L_z$ is decomposed into regions associated with the bulk material

$$L_b = \int_{\{z:\rho(\tau_c,z)\geq\rho_b\}} \mathrm{d}z, \tag{3.5}$$

the craze

$$L_c = \int_{\{z:\rho(\tau_c,z)\leq\rho_c\}} \mathrm{d}z \tag{3.6}$$

and an intermediate length

$$L_i = \int_{\{z:\rho_c<\rho(\tau_c,z)<\rho_b\}} \mathrm{d}z, \tag{3.7}$$

where $\rho(\tau_c, z)$ is the density of a slab $\mathrm{d}z$ at $\tau_c$ and $\rho_b = 0.9a^{-3}$ and $\rho_c = 0.17a^{-3}$ are the bulk and craze density, respectively. The evolution of those craze length components is displayed in Figure 3.6, including additionally snapshots of the craze (created with OVITO [89]) for the three simulations compressed to different $\lambda_{z,u}$-values, and then restrained to the same $\lambda_z = 5$. To further facilitate the understanding of the approach, the area enclosed by the coloured dashed lines in snapshot Figure 3.6(a) defines the three contributions associated with $L_b$, $L_c$ and $L_i$.

In all three simulations, the simulation box consists solely of crazed material ($L_c = L_z$) at $\tau_c = 0$. As the deformation is retracted and the craze unloaded, $L_i$ increases monotonically to $L_z$. In the case of $\lambda_{z,u} = 1$, the high compressive stress eventually leads to a very rapid increase from $L_b = 0$ to $L_b = 1$ (cf. green line around $\tau_c \approx 0.5$ in Figure 3.6(a)). Upon reloading, $L_i$ quickly rises and subsequently $L_b$ as well as $L_i$ continuously decrease. This is associated with the redrawing of fibrils as can be seen in the corresponding animation. The small rise of $L_i$ can be understood as new *active zone* (magenta area shown in the Figure 3.6(a) snapshot), from which fibrils are drawn. In this context, it is important to realize that the decrease in $L_i$ results merely from the normalisation, but is rather a constant active zone length (see Figure 3.B.5 in a non-normalised plot for $L_i$). We also note that it would be misleading to characterise this behaviour as healing of the craze, as neither the temperature nor the simulation time allows for any substantial chain reptation. The response arises rather from the breakage of intermolecular interactions, which were created during $\sigma_z < 0$ (cf. Figure 3.B.3(a)).

This behavior is in strong contrast to $\lambda_{z,u} \geq 2$, where $L_b$ remains zero. While the key feature of $\lambda_{z,u} = 5$ is the retention of the craze morphology seen by smooth transitions

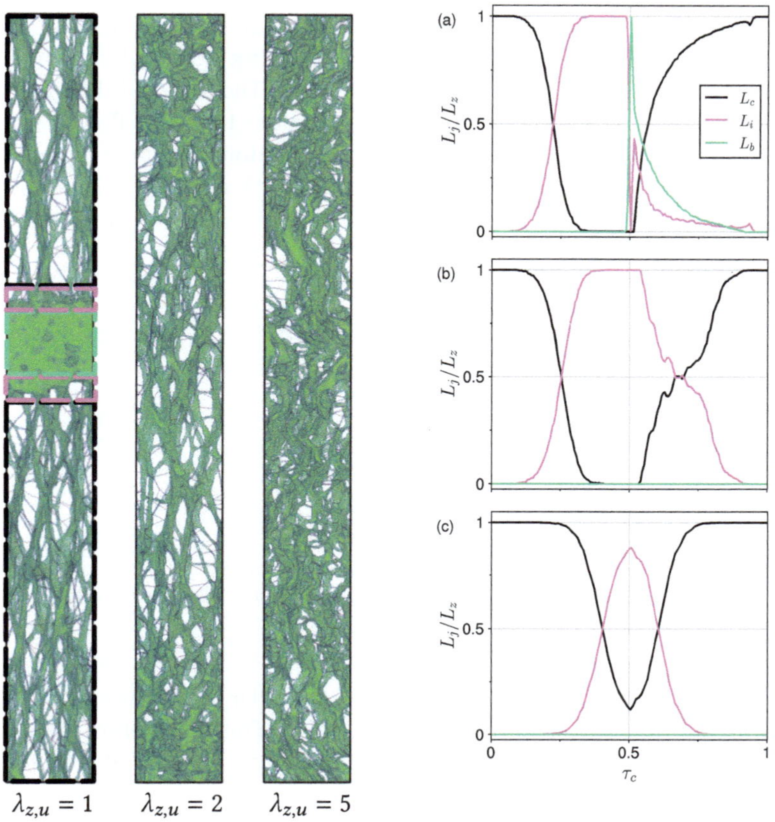


**Figure 3.6.:** Snapshots of crazes during reloading at $\lambda_z = 5$ and corresponding evolution of craze length components throughout cycle for (a) $\lambda_{z,u} = 1$, (b) $\lambda_{z,u} = 2$ and (c) $\lambda_{z,u} = 5$. Colour coding of boxes in snapshot for $\lambda_{z,u} = 1$ indicates three density regions as specified in the text.

between $L_c$ and $L_i$ in Figure 3.6(c), $\lambda_{z,u} = 2$ is more interesting. With the system reloading from $L_i = 1$ at $\tau_c = 0.5$ (Figure 3.6(b)), recavitation does not occur and the breaking of LJ interactions (cf. Figure 3.B.3(b)) results in a relatively constant plateau stress (cf. Figure 3.1(c)). The animation shows an *unfolding* behaviour of the craze, which can bee seen by the evolution around $0.5 < \tau_c < 0.8$ in Figure 3.6(b). This becomes even more evident if the normalisation is omitted and $L_i$ and $L_c$ are plotted separately as shown in Figure 3.B.4.

### 3.3.2. Bulk-craze interaction

The study so far has focused on the cyclic deformation of fully crazed matter, which has been drawn from the isotropic glass. Yet, fibril drawing is a transient process and hence it is also interesting to consider the cyclic response while craze and bulk material coexist.

To investigate the role of the craze length, two configurations are created by deforming the initial glass to $\lambda_{z,max} = 5$ as well as $\lambda_{z,max} = 2$, leading to craze/bulk length ratios at $\lambda_{z,max}$ of $L_c/L_b = 9.8$ and $L_c/L_b = 1.3$, respectively. The protocol then follows the previous protocol by commencing the cyclic loading routine to three different $\lambda_{z,u}$-values, leading to the bulk-craze interaction for the nine simulations shown in Figure 3.7. The deformation has been rescaled by $\varepsilon_z = (\lambda_z - 1)/(\lambda_{z,max} - 1)$, resulting in a collapse of the curves for a given unloading $\varepsilon_{z,u} = (\lambda_{z,u} - 1)/(\lambda_{z,max} - 1)$.

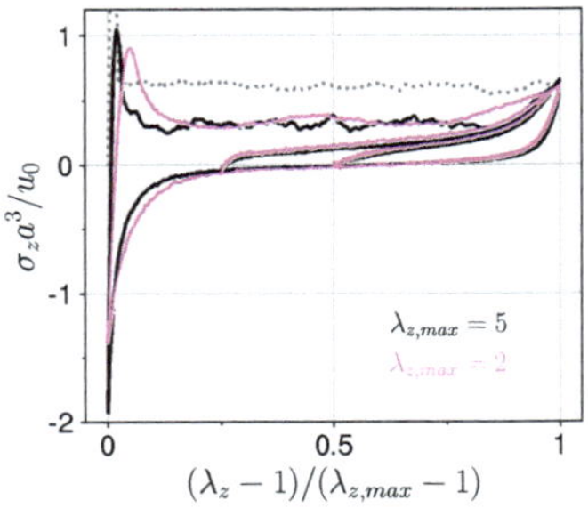


**Figure 3.7.:** Stress-strain curves for two bulk-craze configurations obtained by $\lambda_{z,max} = 5$ (black lines) and $\lambda_{z,max} = 2$ (magenta lines) as well as the initial craze formation (grey dotted line). Each configuration features three unloading levels $\lambda_{z,u}$.

The bulk-craze interaction leads to a response very similar to the pure craze response in Figure 3.1 even for *short* crazes such as $L_c/L_b \approx 1$. This allows for two important conclusions: Firstly, the macro deformation is governed by the craze and the bulk plays a minor role in cyclic deformation. This is attributed to the much higher bulk stiffness $E_b$ compared to the craze stiffness $E_c$. To further elaborate that, the bulk/craze stiffness can be crudely estimated by their respective secant stiffness with respect to $\lambda_z$ while neglecting 3D effects. This yields $E_b/E_c \approx 100$, where $E_b$ is evaluated at $\lambda_z = (1, 1.5)$ and $E_c$ for reloading in regime I at $\lambda_z = (9.5, 10)$ in Figure 3.1(c). It is important to acknowledge that this is the craze stiffness and not the craze fibril stiffness. Secondly, the stress free contraction occurring also for shorter craze fibrils corroborates a negligible bending stiffness and hence the string-like structural response of craze fibrils.

We conclude this section by discussing an estimate of the craze length $\xi_0$ at $\lambda_{z,max}$. This estimate is limited to the case without pore space closure (i.e. $\varepsilon_z \geq 0.25$), which is then motivated by the scaling in Figure 3.7. While drawing is not exhausted, the craze essentially exhibits a layered bulk-craze structure at $\lambda_{z,max}$ in which the deformed craze layer is denoted by $\xi$ and the deformed bulk layer by $\chi$. They differ from $L_c$ and $L_b$ as $\xi$ and $\chi$ are calculated in the following and not given by the MD simulations. Using further continuum micromechanical arguments (for details including finite strains see [56]) yields the kinematic coupling between the macro stretch $\lambda_z$ and the layers $\xi$ and $\chi$ as

$$\lambda_z = \frac{\chi + \xi}{h_0}, \tag{3.8}$$

where $h_0 = L_{z0}$ is the primordial thickness. Furthermore, the unloaded bulk $\chi_0$ and craze length $\xi_0$ are related by conservation of mass during the drawing process, given by

$$\chi_0 = h_0 - \xi_0/\lambda_c, \tag{3.9}$$

where $\lambda_c$ is the extension ratio with respect to the unloaded configuration. To combine (3.8) and (3.9) without proposing constitutive deformation models, we assume small (elastic) deformations of the bulk and craze, i.e. $\chi_0 \approx \chi$ and $\xi_0 \approx \xi$, allowing to derive an estimation for the craze fibril length

$$\xi_0 = h_0(\lambda_{z,max} - 1)\frac{\lambda_c}{\lambda_c - 1} \tag{3.10}$$

at $\lambda_{z,max}$. Applying (3.10) to $\lambda_{z,max} = 5$ and $\lambda_{z,max} = 2$ leads to $\xi_0/\chi_0 = 10$ and $\xi_0/\chi_0 = 1.3$, respectively, which is in good agreement with $L_c/L_b$ values given by the MD simulations.

As a final remark, it is noteworthy that no continuous drawing is observed on the investigated time scales, which would be an indication for viscous drawing or viscous deformation effects of the craze fibril.

## 3.4. Conclusions

Molecular dynamics simulations were used to study the mechanical response of sole fibrillated craze matter and the bulk-craze interaction in glassy polymers under cyclic loading. The maximum loading amplitude $\lambda_{z,max}$ and unloading magnitude $\lambda_{z,u}$ were varied to simulate different bulk-craze compositions and different degrees of pore space closure, respectively. A key finding of this study was that in all investigated cases, i.e. the sole craze and the craze-bulk system, the macro responses exhibited a hysteresis, which was found to be quasi stationary after the first cycle and largely independent of deformation rate and temperature.

Another salient finding was that the hysteresis cannot be simply described by the chain orientation $S_z$, but rather resulted from a complex interplay between constraints imposed by the entanglement network, pore space and pore space closure. Three distinct craze stages were identified to study the driving mechanism in detail: The first stage focused on the highly oriented bonds and craze fibrils. Further deformation led to an accelerated increase in stress $\sigma_z$ (cf. Figure 3.1), which resulted from the stretching of the covalent backbone bonds. Upon unloading the oriented craze, the bond force $\langle F_b \rangle$ quickly relaxed due to the loss in entanglement contact, yielding a rapid drop in $\sigma_z$. The second stage characterised the stress-free contraction during further unloading, in which the animations showed that the craze fibrils undergo a folding motion. The folding motion was characterised on the chain and particle level observables by a decrease in chain orientation and elevated lateral particle movement, respectively. The surrounding pore space is the essential trait for the folding motion to take place as a rigid body motion (i.e. stress-free). The third stage described the reloading response, which exhibited a strong dependency on $\lambda_{z,u}$. It was shown that a key feature is the degree of pore space closure and the necessary intermolecular interaction to enforce it. Complete pore space closure ($\lambda_{z,u} = 1$) was accompanied by a high level of intermolecular stress, which led to a recavitation and redrawing during reloading. If the pore space was maintained throughout the cyclic (e.g. $\lambda_{z,u} = 5$), the reloading response was bi-linear.

The study on the craze-bulk interaction revealed that even *short* craze fibrils, where $L_b \approx L_c$, exhibited the three characteristic stages of sole fibrillated craze matter described above. That is, once a craze initiates and reaches a certain length, the macroscopic response is governed by the craze. This was attributed to the significantly lower craze stiffness (with respect to $\lambda_z$) of $E_b/E_c \approx 100$. It also allows to use continuum micromechanical considerations to estimate the craze-bulk ratio at peak loading.

A further important finding is that there has not been an indication of a macroscopic bending stiffness during unloading, even in the case of short craze fibrils. The folding motion occurs on the craze fibril level which comprises locally stiff craze fibril segments (cf. animations). Therefore, it is likely that the results qualitatively hold for semiflexible bead-spring models, i.e. with a bending potential.[1] While such models primarily yield a higher entanglement density, the crazing process is still dilatant by creating pore space and hence, the mechanisms leading to the stress-free contraction of the craze are still present. Concluding, the macroscopic structural response of craze fibrils during unloading is most accurately described as string-like, despite locally stiff craze fibril segments. This finding is important, since a craze fibril bending stiffness would have different implications on the damage mechanisms during cyclic (fatigue) loading.

The most interesting contribution for future work includes carefully conducted experiments on fibrillated craze matter investigating the response under different loading conditions.

# Appendix to Chapter 3

## 3.A. Animations

Three animations, created with OVITO, are supplemented, featuring a deformation to $\lambda_{z,max} = 10$ and three consecutive loading cycles for

- MS1 (MP4): $\lambda_{z,u} = 1$ where the colour coding displays the increment of the von Mises shear strain invariant. Additionally, the crazing process from the isotropic glass is shown.

[1] A later analysis, which was not published in [55], on semiflexible bead-spring models verified the assumption (cf. Section 3.C).

- MS2 (MP4): $\lambda_{z,u} = 2$ where the colour coding also displays the increment of the von Mises shear strain invariant and
- MS3 (MP4): $\lambda_{z,u} = 5$ where the colour coding displays the lateral particle displacement $u_{xy}$.

The von Mises shear strain is calculated based on the atomic Green-Lagrangian strain tensor and displays here only the nonaffine deformation of the particles. Note, to enhance the visualisation of the otherwise small simulation box, the simulation box is replicated three times in lateral direction, which corresponds to the height in the animations.

## 3.B. Additional data

The additional data presents influence of deformation rate and temperature in Figure 3.B.1,

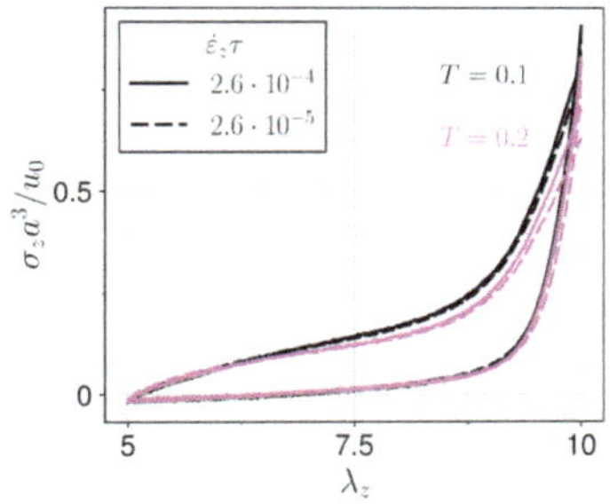


**Figure 3.B.1.:** Influence of deformation rate $\dot{\varepsilon}_z$ and temperature on stress $\sigma_z$ for $\lambda_{z,u} = 5$.

stress evolution for multiple loading cycles in Figure 3.B.2,

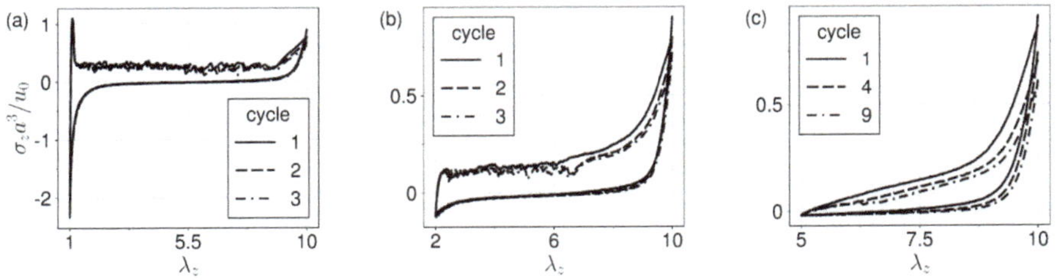


**Figure 3.B.2.:** Stress evolution for multiple consecutive loading cycles for (a) $\lambda_{z,u}$=1, (b) $\lambda_{z,u}$=2 and (c) $\lambda_{z,u}$=5.

stress decomposition for $\lambda_{z,u} = 1$ and $\lambda_{z,u} = 2$ in Figure 3.B.3

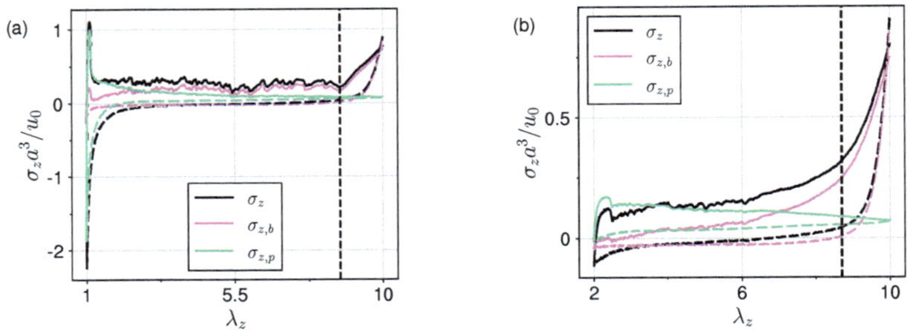


**Figure 3.B.3.:** Stress decomposition for (a) $\lambda_{z,u} = 1$ and (b) $\lambda_{z,u} = 2$ into intermolecular pair $\sigma_{z,p}$ and intramolecular bond $\sigma_{z,b}$ components during unloading (dashed lines) and reloading (solid lines).

evolution of craze length components for $\lambda_{z,u} = 2$ and $\lambda_{z,u} = 1$ in a non-normalised manner in Figure 3.B.4 and Figure 3.B.5, respectively.

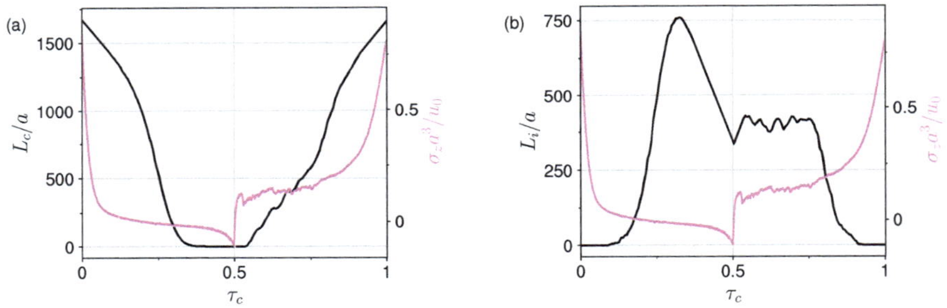


**Figure 3.B.4.:** Evolution of (a) craze length $L_c$ and (b) intermediate length $L_i$ (black) throughout the first cycle for $\lambda_{z,u} = 2$, together with the macro stress $\sigma_z$ (magenta). For $t/T > 0.5$, $L_i$ remains constant while $L_c$ increases indicating the *unfolding* of the craze during reloading.

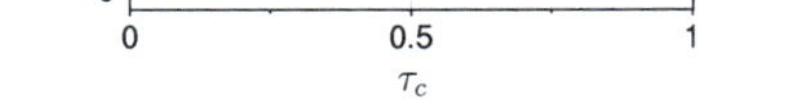


**Figure 3.B.5.:** Evolution of $L_i$ (left axis) and $\sigma_z$ (right axis) for $\lambda_{z,u} = 1$. The relatively constant plateau at $t/T > 0.5$ indicates a constant *active zone* length.

# 3.C. Cyclic craze response of semiflexible polymer glass

Section 3.C is not included in [55] and presents additional results for semiflexible polymer glass. The conclusion that craze fibrils are structurally best described as strings during unloading due to a folding motion was deduced for the flexible bead-spring model in Chapter 3. It was argued that the folding motion occurs at the craze fibril level rather then at the chain level and hence, the conclusions were assumed to similarly hold for semiflexible bead-spring models, incorporating a bending potential. This assumption is investigated in the following.

The semiflexible bead-spring model has the same system size and setup as described in Section 3.2, with the addition of a standard cosine bond bending potential [10]

$$U_{\mathrm{b}}(\theta) = k_b(1 - \cos\theta), \tag{3.11}$$

where $\theta$ is the angle between two consecutive bond vectors along a chain, given by $\cos\theta = (\mathbf{r}_i - \mathbf{r}_{i-1}) \cdot (\mathbf{r}_{i+1} - \mathbf{r}_i)$, where $(\mathbf{r}_i - \mathbf{r}_{i-1})$ and $(\mathbf{r}_{i+1} - \mathbf{r}_i)$ are the unit vectors of the two consecutive bonds in bond direction. The bending potential reduces entanglement length $N_e$ and, consequently, the extension ratio $\lambda_c$. In the model, the bending stiffness is set to $k_b/u_0 = 1.5$, resulting in $N_e \approx 30$ and $\lambda_c \approx 4.4$ (cf. [32]).

The macroscopic stress response is depicted in Figure 3.C.6 for the initial deformation of the isotropic glass (grey line) and the two subsequent loading cycles (black and magenta lines). The deformation protocol follows that outlined in Section 3.2. The results are qualitatively consistent with those observed for the flexible polymer (cf. Chapter 3). In conclusion, the mechanisms analysed for flexible polymers appear to be largely independent of the bending potential. An essential result is the quasi stress-free contraction, which further supports the initial assumption that craze fibrils are most accurately described as string-like structures during unloading.

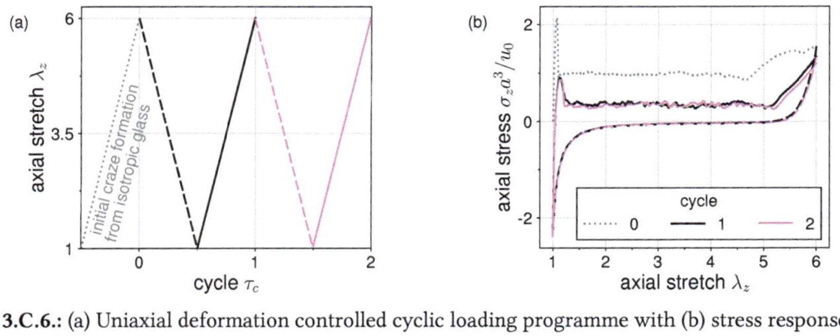


**Figure 3.C.6.:** (a) Uniaxial deformation controlled cyclic loading programme with (b) stress response of sole fibrillated craze matter for two loading cycles.

# 4. Analysis of mode I crack propagation in glassy polymers under cyclic loading using a molecular dynamics informed continuum model for crazing

**This chapter reproduces:***
Laschuetza T and Seelig T. "Analysis of mode I crack propagation in glassy polymers under cyclic loading using a molecular dynamics informed continuum model for crazing". In: *Journal of the Mechanics and Physics of Solids.* 194: 105901, 2025. DOI: `10.1016/j.jmps.2024.105901`

**Abstract:** Craze and crack propagation in glassy polymers under cyclic mode I loading are investigated by employing a recently developed continuum-micromechanical model for crazing. This model accounts for the local morphology change from microvoids to fibrils during craze initiation, viscoplastic drawing of bulk material into fibrils, and viscoelastic creep recovery of the fibrillated craze matter during unloading. To ensure consistency between the bulk and craze model parameters, the material parameters of the craze model are normalised and calibrated based on a hybrid approach integrating experimental findings from the literature and molecular dynamics results. This yields a generic, yet representative glassy polymer response.

In the framework of 2D plane strain finite element simulations, we study brittle as well as ductile glassy polymers and assess the results by drawing comparisons to the experimental and numerical literature. For brittle materials, characterized by a purely elastic bulk behaviour, the model reproduces craze characteristics such as the craze opening contour, the craze length-to-width ratio, a double stress peak at the craze and crack tip, and a non-proportional stress redistribution during loading-unloading cycles. In ductile glassy polymers, the interaction of shear yielding in the bulk and crazing along the ligament is analysed. In particular, shear bands emanate from the crack tip in each loading cycle and arch forward towards the craze. This plastic zone shares resemblance to the so-called *epsilon-shaped deformation zone.* The current simulations capture *normal fatigue crack propagation*, where craze and crack growth occur near the peak load in every cycle and the craze length remains relatively constant across the loading cycles. Moreover, findings from this study suggest that plasticity-induced unloading of the craze adjacent to the crack tip impedes crack growth.

**Keywords:** Craze, Cyclic loading, Micromechanical continuum model, Fracture, Polymeric material, Crack tip plasticity



## 4.1. Introduction

Crack growth in glassy polymers is typically accompanied by crazing in a narrow zone ahead of the crack tip. This damage process involves the formation and coalescence of microvoids, fibrillation of the polymer material in between and drawing of new material into the load-bearing fibrils. By ultimate rupture of the latter the craze zone locally turns into a crack. The energy dissipated in this *process zone* determines the fracture toughness of the material and hence is of practical importance. Crazing may occur as the sole inelastic deformation process in brittle glassy polymers such as polystyrene (PS), or it may take place in conjunction with shear yielding of the surrounding material in more ductile polymers such as polycarbonate (PC). Crazing in the course of crack propagation under monotonic as well as cyclic (e.g. fatigue) loading conditions has been subject of numerous experimental studies; see, e.g., the review articles in [34, 46, 48, 49, 66]. Some key findings in case of cyclic loading are as follows: The fibrillated craze matter displays a pronounced viscoelastic behaviour in terms of deformation and creep recovery during the loading and unloading stages [38, 47]. Compressive stresses at a crack tip upon unloading, which are likely to result from folding (or jamming) of the loose-hanging craze fibrils, are reported e.g. in [70, 79]. Moreover, in the range of low loading amplitudes crack advance may take place not in every cycle but by jumps after remaining stationary for hundreds of fatigue cycles, which is referred to as *discontinuous* or *retarded* fatigue crack growth, e.g. [50, 83], in contrast crack advance in each cycle is understood as *normal* fatigue crack growth. During retarded fatigue crack growth and between successive crack jumps, the thickening of the craze in the first half of its lifetime is primarily due to fibril drawing, whereas in the second half it is predominantly influenced by fibril creep, as reported in [50]. Crack growth in ductile glassy polymers under cyclic loading may exhibit an interesting interaction of crazing and shear yielding which gives rise to the occurrence of a regularly spaced $\varepsilon$-shaped plastic zone accompanying the advancing crack tip [63, 93]. Such an interaction is understood to significantly increase the material's load bearing capacity under cyclic (fatigue) loading conditions [94].

Theoretical-computational studies of craze and crack propagation in glassy polymers have followed various modelling approaches with the majority restricted to monotonic loading conditions. Early studies employed a Barenblatt-Dugdale type representation of the craze zone ahead of an advancing crack tip, e.g. [21, 43, 51]. Other approaches, e.g. [15, 71, 99,

103], aimed at determining the mechanical response of the craze matter by computing the normal stress distribution along a craze in a linear elastic medium from the measured opening displacement profile (craze contour). It has, however, been shown in [104] that in this approach small deviations in the input displacement profile give rise to significant changes in the computed stress. Fatigue crack propagation in polymers was investigated, e.g., in [61] who utilized a phenomenological cohesive zone model and focused on the overall response of the fracture process in terms of Paris' law. A more advanced cohesive zone model that incorporates physical details of the crazing process such as initiation and drawing of material into fibrils until ultimate breakdown has been developed by Van der Giessen and co-workers [97]. Utilizing this model, mode I craze and crack propagation under monotonic loading was studied in brittle (elastic) glassy polymers by [96] and in ductile glassy polymers by [26] who particularly analysed the interaction of shear yielding and crazing. The present work is aimed as a continuation of these studies by focusing on cyclic loading conditions.

Modelling the mechanical response of the fibrillated craze matter under cyclic loading is more involved than under monotonic loading where it is essentially governed by a (rate-dependent) drawing stress. For instance, the structural response of the fibrillated craze matter during unloading and reloading is ambiguous and hence is its proper representation. This includes relaxation and creep recovery during unloading stages when fibril drawing is interrupted, the occurrence of compressive stresses due to fibril jamming as well as cyclic damage accumulation. In order to capture these aspects, a novel continuum-micromechanical model for crazing has recently been developed in [56]. In a finite strain setting, the model builds upon micromechanical considerations by Boyce and co-workers [82, 84] to distinguish between the already fibrillated and the not yet fibrillated bulk portions of a *representative craze element* and their conversion in the course of the crazing process. In the present work, this model serves as a *traction separation law* in the process zone ahead of a mode I crack initiating and advancing in a glassy polymer under cyclic loading. However, while the model developed in [56] aims to capture important physics of the cyclic crazing process through its micromechanical basis, it still suffers from inconclusive knowledge about the structural response of the craze matter and insufficient calibration methodologies to estimate an adequate material parameter scope. This is an issue as insight into both aspects is difficult to obtain from experiments. However, molecular dynamics simulations might provide remedy. This type of *bottom-up* computational investigation has in the last decades significantly contributed to the understanding of the crazing process and its dependence on macro-molecular characteristics such as the entanglement density; see, e.g. [32, 76, 100]. While these and other studies so far have focused on monotonic loading conditions, molecular dynamics simulations of the crazing process under cyclic loading have only recently been conducted in [55]. Insights from that study verify the structural assumption of string-like craze fibrils in [56]. In this work, we utilise and extend results from [55] to enrich the original crazing model from [56] by adding additional features, e.g. fibril jamming, and by using molecular dynamics simulations to calibrate material parameters, which are not accessible from experiments. The model is then used to study craze and crack growth under cyclic mode I loading, while comparisons to the aforementioned experiments are drawn.

The present work is organized as follows: In Section 4.2 the micromechanical crazing model is presented and the methodology of the parameter calibration is discussed as well as some numerical aspects. Details on the molecular dynamics based calibration are provided in Appendix 4.A. Finite element simulations of craze and crack growth in glassy polymers under cyclic mode I loading are subject of Sections 4.3 and 4.4. Section 4.3, which also includes the set-up of the boundary value problem and details on the computational treatment, focuses on brittle glassy polymers where the bulk material surrounding the craze and crack is considered linear elastic. The interaction of crazing and shear yielding during crack growth under cyclic loading in ductile glassy polymers is studied in Section 4.4. The constitutive model utilized to describe finite strain viscoplastic shear yielding in the bulk material is based on the well-known model by Boyce and co-workers [17] in a setting which is briefly presented in Appendix 4.B.

The notation throughout this paper makes use of the standard symbolic bold face representation of vectors and second order tensors as well as the Cartesian index notation of their components.

## 4.2. Continuum-micromechanical model for crazing

The main topic of this paper is the computational investigation of craze and crack growth in glassy polymers under cyclic mode I loading (cf. Figure 4.1 top). Therefore, a continuum-micromechanical model for the cyclic craze response developed in [56] is utilized in the present study to provide a macro-scale traction-separation law. This model considers a *representative craze element* where an elementary volume of initial bulk material (Figure 4.1(a)) transitions into crazed material with a current unloaded and loaded configuration sketched in Figure 4.1(b) and (c), respectively. Details of the model are briefly summarised in the following three subsections, while details on the parameter calibration, the model response and the numerical treatment are provided in Subsections 4.2.4, 4.2.5 and 4.2.6, respectively.

### 4.2.1. Micromechanical model

Prior to craze initiation, the craze element consists of bulk material of the initial primordial thickness $h_0$. At some time $t$ after craze initiation, the craze element comprises layers of fibrillated matter with the current unloaded length $\xi_0(t)$ and bulk material with the current unloaded length $\chi_0(t)$ as indicated in Figure 4.1(b); the corresponding lengths in the currently loaded configuration are denoted by $\xi(t)$ and $\chi(t)$, respectively (cf. Figure 4.1(c)).

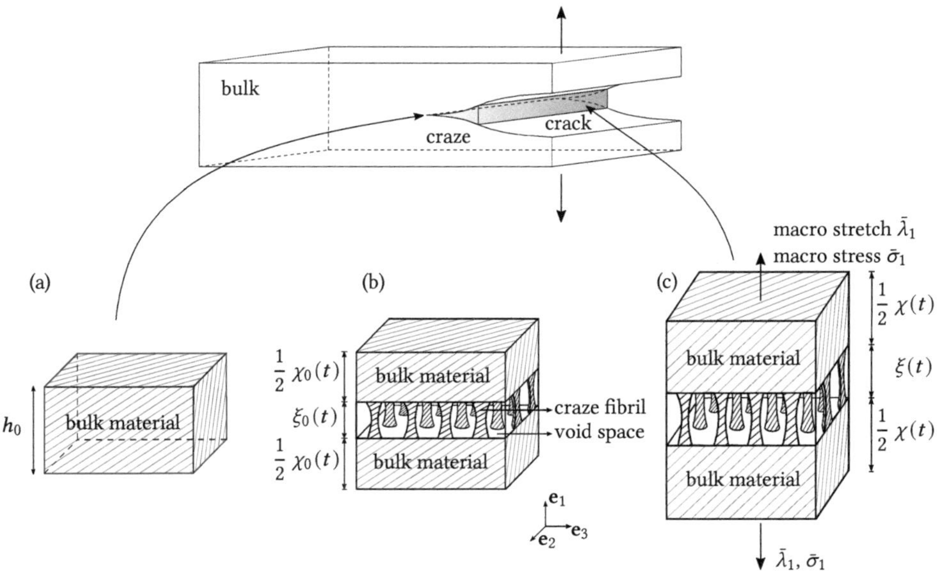


**Figure 4.1.:** Schematic of mode I craze and crack growth with different stages of crazing process; (a) prior to craze initiation, (b) unloaded state after craze initiation and (c) loaded state after craze initiation.

The complex craze microstructure is idealised by neglecting cross-tie fibrils and considers only string-like fibrils in the direction $\mathbf{e}_1$ of the maximum principal stress (Figure 4.1). Mass conservation links the thicknesses of the two phases as

$$h_0 = \chi_0 + \frac{\xi_0}{\lambda_c}, \tag{4.1}$$

where $\lambda_c$ denotes the fibril extension ratio, defined here as the ratio of the constant bulk density $\rho_b$ to the density $\rho_c$ of the fibrillated craze matter:

$$\lambda_c = \frac{\rho_b}{\rho_c}. \tag{4.2}$$

In order to capture the change of $\rho_c$ (and hence $\lambda_c$) due to the morphological transition during craze formation from isolated voids to an interconnected void space with isolated fibrils, the extension ratio $\lambda_c$ is taken to evolve with the unloaded fibril length $\xi_0$ from $\lambda_c = 1$ (bulk value) to a saturation value $\lambda_c = \lambda_c^*$ that corresponds to mature fibrils. The transition of $\lambda_c(\xi_0)$ is depicted in Figure 4.2 where $\xi_{0,\max} = \lambda_c^* h_0$ is the unloaded fibril length at rupture and $\alpha$ defines its fraction upon which mature fibrils exist.

The stress state in the fibrils is taken as homogeneous and uniaxial, yielding

$$\boldsymbol{\sigma}^f = \sigma^f \mathbf{e}_1 \otimes \mathbf{e}_1, \tag{4.3}$$

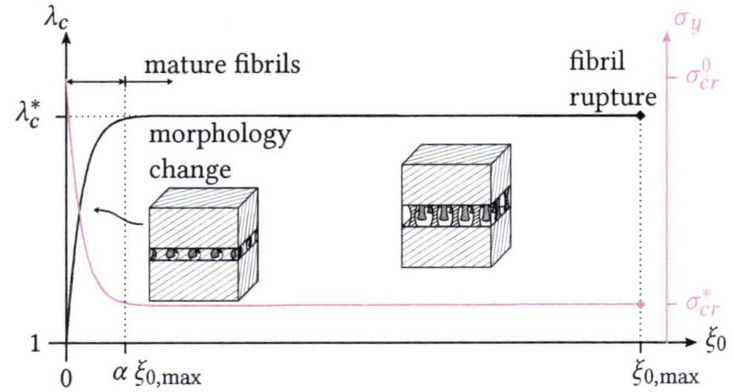


**Figure 4.2.:** Phenomenological ansatz to account for effect of morphology change on evolution of extension ratio $\lambda_c$ as well as drawing resistance $\sigma_y$ with fibril length $\xi_0$. Both relations are modelled by similar exponential relations that approximately connect the limit states. Sketches serve to facilitate interpretation of morphology stages.

where $\sigma^f$ is the Cauchy fibril stress. Likewise, the stress state in the bulk portion of the craze element

$$\boldsymbol{\sigma}^b = \sum_{i=1}^{3} \sigma_i^b \, \mathbf{e}_i \otimes \mathbf{e}_i, \tag{4.4}$$

is assumed to be homogeneous. Homogenisation of the layered bulk-craze structure of the element yields through the *rule of mixtures* the macroscopic Cauchy stress $\bar{\boldsymbol{\sigma}}$ as

$$\bar{\boldsymbol{\sigma}} = \sigma_1^b \, \mathbf{e}_1 \otimes \mathbf{e}_1 + \frac{\chi}{\chi + \xi} \sum_{i=2}^{3} \sigma_i^b \, \mathbf{e}_i \otimes \mathbf{e}_i. \tag{4.5}$$

The overall deformation of the craze element is described by the macroscopic stretch tensor

$$\bar{\boldsymbol{\lambda}} = \sum_{i=1}^{3} \bar{\lambda}_i \, \mathbf{e}_i \otimes \mathbf{e}_i, \tag{4.6}$$

which is related to the bulk stretch tensor

$$\boldsymbol{\lambda}^b = \sum_{i=1}^{3} \lambda_i^b \, \mathbf{e}_i \otimes \mathbf{e}_i \tag{4.7}$$

and the fibril stretch in $\mathbf{e}_1$-direction $\lambda^f = \xi/\xi_0$ by

$$\bar{\lambda}_1 = \frac{\xi + \chi}{h_0} = \lambda_1^b \frac{\chi_0}{h_0} + \lambda^f \frac{\xi_0}{h_0}, \qquad \bar{\lambda}_2 = \lambda_2^b, \qquad \bar{\lambda}_3 = \lambda_3^b. \tag{4.8}$$

Force equilibrium in $\mathbf{e}_1$-direction and the assumption of isochoric fibril deformation yields (for details see eqs. (10)-(14) in [56])

$$\sigma_1^b = \frac{\sigma^f}{\lambda^f \, \lambda_c \, \lambda_2^b \, \lambda_3^b}, \tag{4.9}$$

which by virtue of (4.5) provides a coupling between the fibril stress and the macroscopic stress in $\mathbf{e}_1$-direction.

The fibril's string-like microstructure suggests that its load-bearing capacity is essentially limited to tension:

$$\sigma^f \geq 0. \tag{4.10}$$

The notion of string-like fibrils which are *loose hanging* when traction free was corroborated by recent molecular dynamics simulations [55], which are also discussed in Appendix 4.A. Additionally, the study revealed a macroscopic compressive stress due to fibril jamming prior to reaching macroscopic compressive deformation, i.e. $\bar{\sigma}_1 < 0$ while $\bar{\lambda}_1 > 1$, which has also been computed from experimentally measured craze profiles (cf., e.g., [15]). This is incorporated in the present craze model via the ratio

$$\hat{\lambda}_1 = \frac{\xi_0 + \chi_0}{h_0}, \tag{4.11}$$

from which the logarithmic Hencky strain

$$\hat{\varepsilon}_1 = \ln \hat{\lambda}_1 \tag{4.12}$$

can be computed, which is utilized below in (4.17). Fibril jamming is then considered for

$$\bar{\varepsilon}_1 = \ln \bar{\lambda}_1 < \beta_c \, \hat{\varepsilon}_1 \tag{4.13}$$

where $\beta_c$ is a material parameter that controls the onset of fibril jamming. The craze element's compressive response is modelled by the bulk material.

### 4.2.2. Craze initiation, fibril drawing and breakdown

Although various more involved craze initiation criteria exist in the literature, e.g. accounting for the effect of hydrostatic stress, craze initiation in the present work is simply associated with a critical normal stress

$$\sigma_1^b = \sigma_{cr}^0. \tag{4.14}$$

Upon craze initiation, fibril drawing is described by the flow rule

$$\dot{\xi}_0 = \frac{h_0}{\eta_d} \left\langle \sigma_1^b - \sigma_y \right\rangle \quad \geq 0, \tag{4.15}$$

which makes use of the standard notation for viscoplastic models with the Macaulay bracket $\langle ... \rangle$ and where $\eta_d$ is the drawing viscosity. The drawing resistance $\sigma_y$ is taken to decline from an initial value $\sigma_{cr}^0$ to a saturation value $\sigma_{cr}^*$ in order to model the change in stress state in the course of the morphology change between craze initiation by cavitation and fibril drawing (see Figure 4.2). This has a similar effect as a craze initiation criterion involving hydrostatic stress as employed, e.g., in [26, 27, 96]. The corresponding relation is depicted by the magenta line in Figure 4.2.

Fibril breakdown is taken here to occur upon complete consumption of the primordial thickness $h_0$, i.e. at $\chi_0 = 0$, which yields

$$\xi_{0,max} = \lambda_c^* \, h_0. \tag{4.16}$$

### 4.2.3. Constitutive equations of bulk and craze layer

The material behaviour in the bulk portion of the craze element is given by

$$\boldsymbol{\sigma}^b = 2\mu_b \left( \ln \boldsymbol{\lambda}^b + \frac{\nu_b}{1 - 2\nu_b} \operatorname{tr} \left[ \ln \boldsymbol{\lambda}^b \right] \mathbf{1} \right), \tag{4.17}$$

where $\mu_b$ and $\nu_b$ denote the bulk shear modulus and Poisson's ratio, respectively, and by virtue of (4.13) $\ln \lambda_1^b = \bar{\varepsilon}_1 - \beta_c \, \hat{\varepsilon}_1$ to account for fibril jamming.

The craze fibrils are modelled as viscoelastic in a three-dimensional finite strain setting. The fibril deformation is for simplicity considered to be isochoric, yielding the fibril stretch tensor $\boldsymbol{\lambda}^f$ as

$$\boldsymbol{\lambda}^f = \lambda^f \mathbf{e}_1 \otimes \mathbf{e}_1 + \sum_{i=2}^{3} \frac{1}{\sqrt{\lambda^f}} \mathbf{e}_i \otimes \mathbf{e}_i. \tag{4.18}$$

In accordance with the viscoelastic model shown in Figure 4.3, the fibril stretch is multiplicatively decomposed

$$\lambda^f = \lambda^e \, \lambda^v \tag{4.19}$$

into an elastic $\lambda^e$ and viscous contribution $\lambda^v$, while the fibril stress is additively given by

$$\sigma^f = \sigma^f_{\mathrm{EC}} + \sigma^f_{\mathrm{NH}}. \tag{4.20}$$

The network stress $\sigma^f_{\mathrm{EC}}$ is modelled via the incompressible eight-chain model by Arruda and Boyce [8]

$$\sigma^f_{\mathrm{EC}} = \frac{\mu_{f,\mathrm{EC}}}{\lambda_C} \frac{\mathcal{L}^{-1}(\lambda_C/\lambda_L)}{\mathcal{L}^{-1}(1/\lambda_L)} \left( \lambda^{f\,2} - \frac{1}{\lambda^f} \right), \tag{4.21}$$

with the inverse Langevin function $\mathcal{L}^{-1}(x)$ replaced by the Padé approximation [20]

$$\mathcal{L}^{-1}(x) = x \frac{3 - x^2}{1 - x^2} \tag{4.22}$$

and the mean chain stretch $\lambda_C$ simplifies under uniaxial stress and isochoric deformation to

$$\lambda_C = \sqrt{\frac{\operatorname{tr}\left[\boldsymbol{\lambda}^{f\,2}\right]}{3}} = \sqrt{\frac{1}{3} \left( \lambda^{f\,2} + \frac{2}{\lambda^f} \right)}. \tag{4.23}$$

Equation (4.21) involves two material parameters, namely the initial shear modulus $\mu_{f,\text{EC}}$ and the limit stretch $\lambda_L$ corresponding to the entanglement density of the glassy polymer. An additional nonlinear spring in series to the dashpot (Figure 4.3, element B) is included to enable a realistic response upon instantaneous loading. The stress is given by an incompressible neo-Hookean material model

$$\sigma^f_{\text{NH}} = \mu_{f,\text{NH}} \left( \lambda^{e\,2} - \frac{1}{\lambda^e} \right), \tag{4.24}$$

which introduces a second shear modulus $\mu_{f,\text{NH}}$ as a material parameter. The viscous fibril deformation is described by the flow rule

$$\dot{\lambda}^v = \frac{2}{3\,\eta_c} \sigma^f_{\text{NH}} \lambda^v. \tag{4.25}$$

Figure 4.3 illustrates the arrangement of rheological models of viscoelastic fibril deformation and viscoplastic fibril drawing. For further details on the craze model and particularly a motivation and discussion of various constitutive assumptions, e.g. the simplified linear flow models with constant viscosities in (4.15) and (4.25), the reader is referred to [56].

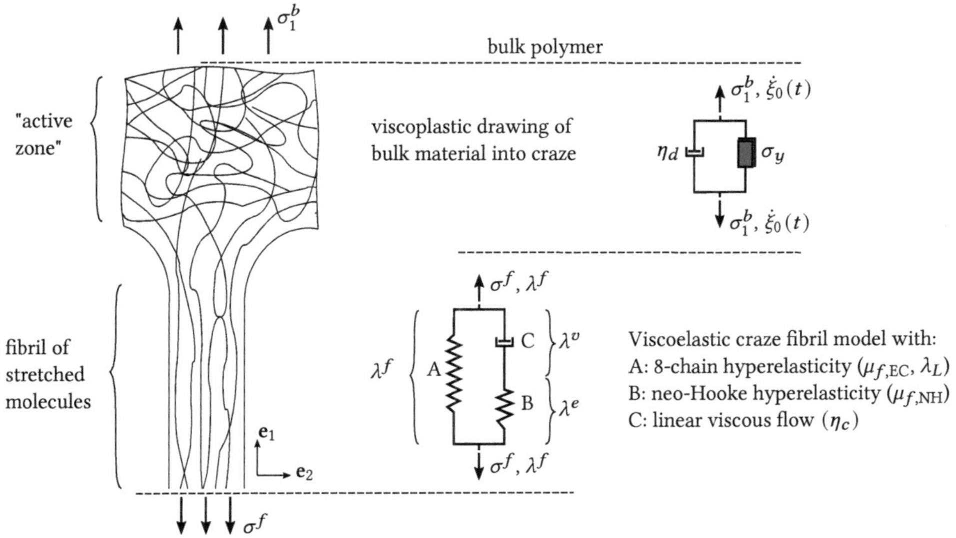


**Figure 4.3.:** Fibril model comprising viscoplastic drawing from the active zone and viscoelastic fibril deformation.

## 4.2.4. Remarks on parameter calibration

We aim to obtain a generic, yet representative craze response for glassy polymers. This presents the challenge that craze and bulk parameter scopes are not independent as they should describe the same (generic) material. Unlike the bulk parameters, which can be calibrated from experiments, the craze parameter calibration is difficult due to the small length scale. We address this issue in several steps to obtain a physically plausible parameter scope of the craze parameters:

First, we take bulk model parameters from the literature (cf. Appendix 4.B), which are representative for glassy polymers. This yields the shear modulus $\mu_b$ and the Poisson's ratio $\nu_b$ for an elastic bulk material (cf. Section 4.3) and additionally the initial shear yield strength $s_0$ for an inelastic bulk material (cf. Section 4.4).

Second, we normalise the craze parameters with $\mu_b$ and $s_0$ where applicable. With the known bulk parameters, the craze model encompasses ten additional constants as well as the primordial thickness $h_0$. We normalise the two characteristic times $\tau_d = \eta_d/\mu_b$ and $\tau_c = \eta_c/\mu_b$, which describe the fibril drawing and the fibril creep behaviour of the craze element, respectively, with a characteristic loading time $T_0$ into the two dimensionless parameters $\tau_d/T_0$ and $\tau_c/T_0$. The geometry of the boundary value problem in Section 4.3 is scaled with $h_0$ and the primordial thickness is set to $2h_0 = 1\mu\text{m}$, which is comparable to previous continuum modelling [82]. The primordial thickness can be understood as a material property which is determined by the (measurable) extension ratio and the (measurable) fibril length (craze opening displacement) at failure ($\chi_0 = 0$) according to (4.1).

Third, we use experimental and numerical findings from the literature to deduce representative values for the extension ratio $\lambda_c^*$, the craze initiation stress $\sigma_{cr}^0$, the saturation drawing stress $\sigma_{cr}^*$ and the two viscosities $\eta_d$ and $\eta_c$: The extension ratio $\lambda_c^*$ has been extensively studied in experiments and $\lambda_c^* \approx 2$ is representative of the bulk parameters in Table 4.B.1 (cf., e.g., [24]). Craze initiation is taken to occur at a representative stress magnitude for glassy polymers, given by $\sigma_y/\mu_f = 0.11 \approx 1.2s_0/\mu_b$. We set the saturation drawing stress to $\sigma_{cr}^* = 2\sigma_{cr}^0/3$, which corresponds to $\sigma_{cr}^* \approx 0.8s_0$ as used in [26]. This facilitates later comparisons under monotonic loading. With set $\sigma_{cr}^0$, the fibril drawing viscosity $\eta_d$ significantly influences the crack propagation speed and hence, the macroscopic stress in the far field. Linear elastic fracture mechanics is used to estimate a far field stress as experimentally observed for the mode I boundary value problem outlined in Section 4.3. The considered range of values for $\tau_d/T_0$ and $\tau_c/T_0$ is given in Subsection 4.3.1 after the boundary value problem has been presented.

Fourth, we employ molecular dynamics (MD) simulations to obtain the parameter scope for the remaining parameters, including the elastic fibril properties $\mu_{f,\text{EC}}$ and $\mu_{f,\text{NH}}$, the limit stretch $\lambda_L$, the parameter $\beta_c$ controlling fibril jamming and the fraction $\alpha$ upon which mature fibrils exist. MD simulations complement well the micromechanical approach by providing insight into microscale mechanisms that are otherwise challenging to access experimentally due to their small length scale. The details of the non-standard molecular dynamics calibration are presented in Appendix 4.A, with key considerations summarised as follows: The elastic part of the constitutive craze fibril model (cf. Figure 4.3) is fitted to the stress-strain response of the MD simulations, yielding $\mu_{f,\text{EC}}$, $\mu_{f,\text{NH}}$ and $\lambda_L$. The parameter scope is further assessed by normalising the quantities with the MD bulk stiffness $\mu_b$, leading to $\mu_b/\mu_f \approx 10$ and $\mu_{f,NH}/\mu_f \approx 0.5$, where $\mu_f = \mu_{f,\text{EC}} + \mu_{f,\text{NH}}$. According to [55], the fibril length in the MD simulations is closely linked to the deformation. Combined with (4.13), this correlation is used to determine the onset of fibril jamming as function of the craze fibril length $\xi_0$, resulting in $\beta_c \approx 0.15$. Similar to the idea of the morphology change depicted in Figure 4.2, the MD simulations allow to determine the instant after cavitation upon which the mature craze density prevails, leading to $\alpha \approx 0.07$.

Finally, we investigate different parameter combinations to study their influence on the (generic) response.

### 4.2.5. Response of calibrated craze model

To study later the effect of $\mu_b/\mu_f$ and $\lambda_L$, three materials are selected as shown in Table 4.1. Material 1 utilises $\mu_b/\mu_f = 1$ and a relatively high limit stretch $\lambda_L$. By contrast, material 2 takes $\mu_b/\mu_f = 10$ according to the MD findings and material 3 additionally is based on a low $\lambda_L$ corresponding to the MD results.

**Table 4.1.:** Material parameter sets of craze model, where $s_0$ and $\mu_b$ are given by the bulk material in Table 4.B.1.

| set | $\mu_b/\mu_f$ | $\nu_b$ | $\mu_{f,NH}/\mu_f$ | $\sigma_y/\mu_f$ | $\sigma_{cr}^0/\sigma_{cr}^*$ | $\lambda_c^*$ | $\alpha$ | $\lambda_L$ | $\beta_c$ | $2h_0$ [µm] |
|---|---|---|---|---|---|---|---|---|---|---|
| 1 | 1 | 0.38 | 0.5 | $1.2s_0/\mu_b$ | 1.5 | 2 | 0.07 | 2 | 0.15 | 1 |
| 2 | 10 | 0.38 | 0.5 | $1.2s_0/\mu_b$ | 1.5 | 2 | 0.07 | 2 | 0.15 | 1 |
| 3 | 10 | 0.38 | 0.5 | $1.2s_0/\mu_b$ | 1.5 | 2 | 0.07 | 1.2 | 0.15 | 1 |

Following the study in [56], the evaluation of the three materials under uniaxial deformation in a strain controlled bilinear cyclic loading programme is depicted in Figure 4.4. The characteristic loading time $T_0$ is set to the reciprocal of the overall deformation rate $T_0 = 1/\dot{\bar{\lambda}}_1$. All three materials exhibit qualitatively similar traits: an initial peak stress followed by a transition to a constant drawing stress plateau, while during unloading fibrils are stress-free and loose hanging until fibril jamming occurs. Notably during reloading in the 2nd cycle, material 1 shows the most pronounced hysteresis and fibril creep recovery

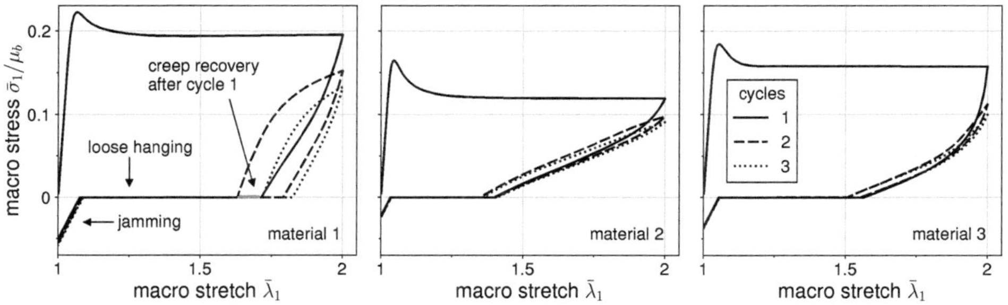


**Figure 4.4.:** Uniaxial strain response of continuum-micromechanical craze model for three consecutive loading cycles and materials outlined in Table 4.1. ($\tau_c/T_0 = 0.1$, $\tau_d/T_0 = 0.1$)

while the fibrils are stress-free, which is represented by the offset between unloading and subsequent reloading in the next cycle (cf. magenta marking in Figure 4.4 for material 1). Except for the curvature of the unloading-reloading curves, material 2 and 3 behave very similar during cyclic loading. We conclude that the change in stiffness $\mu_b/\mu_f$ exhibits the most pronounced influence on the results and thus, we use material 2 as the base case in the subsequent study unless indicated otherwise.

### 4.2.6. Numerical aspects

The craze model is defined by a system of differential algebraic equations and implemented as a user material subroutine in the finite element programme ABAQUS [2]. To compute the tangent directly within the Newton-Raphson scheme, the FORTRAN implementation employs *dual number automatic differentiation.* To that end, we extended the dual number automatic differentiation tool developed by Yu and Blair [107] to accommodate the specific operations required in the implementation. Both, the craze model described in Section 4.2 and the bulk model for shear yielding outlined in Appendix 4.B, are implemented with this methodology. This computational approach has been successfully deployed on the high-performance computing platform *bWUniCluster 2.0*, where both material models have been simultaneously applied.

## 4.3. Craze and crack growth in brittle glassy polymers

### 4.3.1. Model set-up and computational aspects

The 2D plane strain boundary value problem (BVP) of a rectangular plate with an edge crack of initial length $a_0$ subjected to mode I loading is depicted in Figure 4.5. Loading is imposed in terms of a displacement controlled bilinear cyclic loading programme $u_y(t, T, R_u)$ with time $t$, period $T$, displacement velocity $\dot{u}_y$ and load ratio $R_u = u_{y,min}/u_{y,max}$ as sketched in Figure 4.5(b). The notch tip radius $r_t = 50h_0$ (see Figure 4.5(c)) is introduced to alleviate mesh distortions and the $r_t$ magnitude is comparable to [26, 27]. Mode I symmetry is exploited and one layer of craze elements with initial thickness $h_0$ is placed along the ligament indicated by the red line in Figure 4.5. All length dimensions scale with the primordial thickness $h_0$. The BVP is solved with the commercial finite element (FE) software ABAQUS/EXPLICIT [2] to properly capture dynamic effects during potentially unstable crack growth. In contrast to various other studies, e.g. [26, 96], the occurrence of dynamic effects also has led us to avoid a small scale yielding BVP. Instead the entire specimen is modelled as shown in Figure 4.5(a). Both, craze and bulk material are implemented as user material subroutines and craze elements are removed once the failure criterion of a critical craze thickness $\xi_0 = \xi_{0,\max}$ is met (cf. (4.16)). Elements along the ligament are quadratic in shape with a dimension of $h_0$. The ratio $r_t/h_0 = 50$ provides an indication of

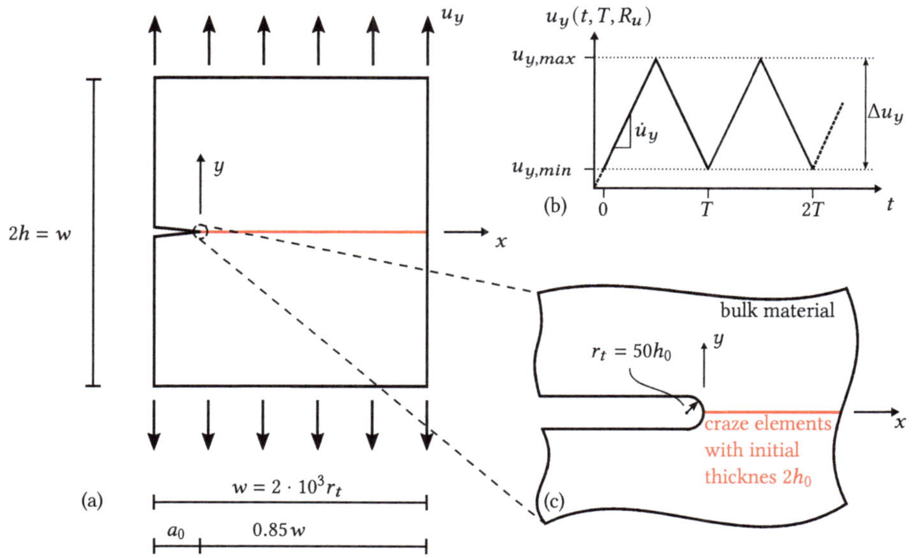


**Figure 4.5.:** (a) Mode I boundary value problem with (b) displacement controlled cyclic loading history and (c) detail of the round crack tip.

the mesh resolution.[1] Since this type of BVP is prone to hourglassing, the finite elements are fully integrated.

The maximum displacement $u_{y,max}$ (Figure 4.5(b)) is chosen to yield a realistic far field stress as observed in experiments. With inelasticity limited to a small region around the crack tip, linear elastic fracture mechanics arguments are employed to estimate $\sigma_y$ in the far field. For the current BVP, the stress intensity factor is $K_I = F\sigma_y\sqrt{\pi a_0}$ with a shape factor $F \approx 0.9$ according to table C10.15 in [29]. We consider two cases: one for an elastic and one for an inelastic bulk material. In the case of a purely elastic bulk response, the fracture toughness of glassy polymers typically ranges around $K_{Ic} \approx 1\text{MPa}\sqrt{\text{m}}$. This yields a realistic far field stress $\sigma_y \approx 7.5\text{MPa}$. The magnitude for more ductile glassy polymers is deduced from experiments where both crazing and shear yielding take place. According to Takemori [94], crack tip plastic zones involving a strong interaction of both mechanisms in PC occur in the range of up to $\Delta K_I = 1.3\text{MPa}\sqrt{\text{m}}$ for $R = F_{min}/F_{max} = 0.1$. This leads to $\sigma_y \approx 10.5\text{MPa}$ as far field stress.

The fibril drawing viscosity $\eta_d$ is chosen so that crack initiation does not take place in the first cycle while, on the other hand, avoiding a fatigue loading regime where thousands of cycles are necessary. To compare varying loading periods $T$ due to, for instance, different $R_u$, the reciprocal of the strain rate in the far field is used as characteristic time $T_0 = h/\dot{u}_y$. This results in $\tau_d/T_0 \approx 10^{-4}$. We remark that the small value arises from the choice of the

[1] The FE mesh is visible in Figure 4.6, showing the situation at some instant after crack initiation.

far field velocity. Rescaling with the ratio of specimen height and primordial thickness $h/h_0 = 5 \cdot 10^4$ yields $\tau_d/T_0 \approx 5$, which is more comparable to the values given in Figure 4.4. However, since the far field velocity is more representative in experiments, this rescaling is omitted. As a normalised measure of the loading magnitude, the far field strain, given by $\varepsilon_y = u_y/h$, is introduced.

In the following, the evaluation mainly focuses on quantities along the ligament. The craze length $l_c$ is computed as the longest contiguous craze segment. That is, $l_c$ neither includes craze sections separated by a crack nor by un-crazed elements. This method helps to focus on the evolution of the most dominant, i.e. longest, craze, which is important since crack initiation may not occur at the notch tip. On the other hand, the crack length $a$ is calculated as the cumulative sum of all cracks, i.e. including separated crack sections.

### 4.3.2. Numerical results

We first study the situation of a purely linear elastic bulk material with craze material 2 (cf. Table 4.1) and corresponding elastic bulk parameters $\mu_b$ and $\nu_b$. Cyclic loading with $R_u = 0$ and $\varepsilon_{y,max} = 2.5 \cdot 10^{-3}$ is considered, resulting in an average far field stress of $\sigma_y \approx 7.6$MPa. If not indicated differently, the normalised drawing and fibril creep viscosities are taken as $\tau_d/T_0 = 5 \cdot 10^{-4}$ and $\tau_c/T_0 = 10^{-4}$, respectively.

Figure 4.6 presents a snapshot of the vertical stress field $\sigma_y$ along with the finite element mesh close to the notch tip at peak load in the 31$^{st}$ cycle some time after crack initiation. As annotated in Figure 4.6, the stress distribution exhibits a double peak at the crack tip and at the craze tip. The craze extends between those peaks, while elements on the right-hand side of the craze tip have not yet initiated.

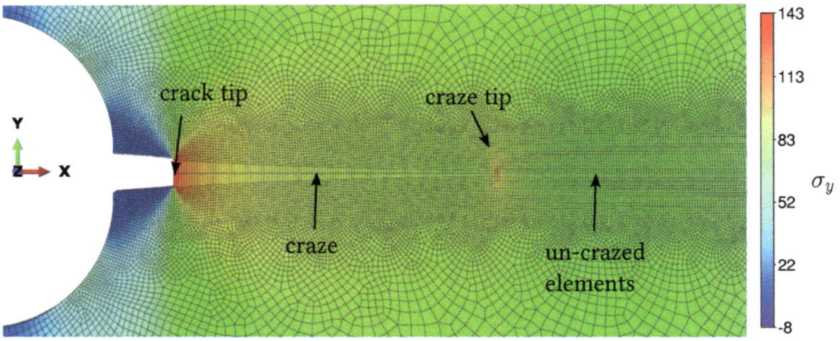


**Figure 4.6.:** Stress field $\sigma_y$ at the notch tip at peak load in the 31$^{st}$ cycle.

The stress distribution and the craze opening contour in terms of the fibril length are shown at four instants throughout the 31$^{st}$ cycle in Figure 4.7 during loading (black line), at peak load (magenta line), during unloading (green line) and at overall zero displacement (orange line). As the elements on the right-hand side of the craze tip have not yet initiated, they still exhibit elastic bulk behaviour, resulting in a quasi-identical stress distribution in stage 1 and stage 3 (Figure 4.7(a)). The double stress peak is visible in instant 2, where the stress at the crack tip exceeds the stress at the craze tip. In contrast, in stages 1 and 3 the stress at the craze tip is higher than at the crack tip. Moreover, the stress at the crack tip during loading (black line) is higher than during unloading (green line), despite the smaller craze width (cf. Figure 4.7(b)). This effect is attributed to the elongation of the craze fibrils due to drawing and creep deformation throughout the loading cycle. The relatively constant shift in the craze contour between stage 1 and 3 (Figure 4.7(b)) suggests uniform drawing along the craze. In contrast, creep deformation depends on the current (relaxed) fibril length $\xi_0$ and becomes more dominant for longer fibrils (cf. [56]), as prevailing at the crack tip. Consequently, the deviation in crack tip stress between instant 1 and 3 is attributed to the fibril creep viscosity. Instant 4 displays the influence of fibril jamming, which is again most pronounced at the crack tip where $\xi_0$ reaches its maximum value. The slight horizontal shift of the stress curves throughout the cycle indicates the small amount of crack propagation.

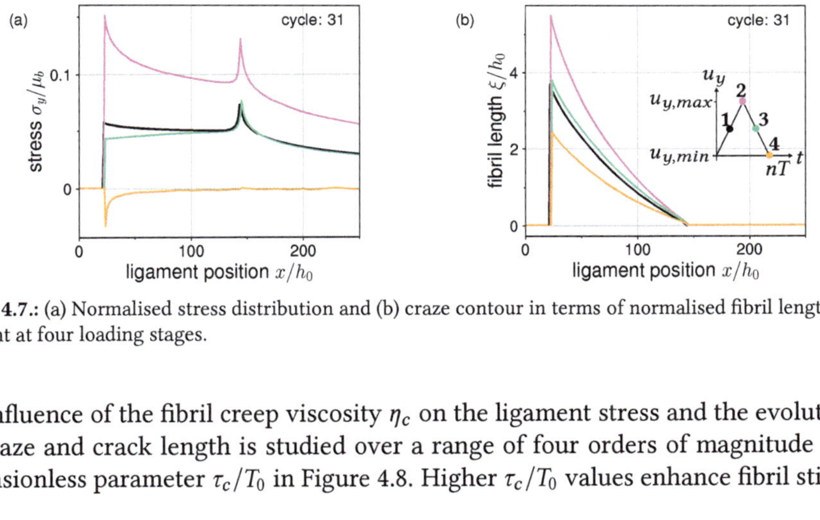


**Figure 4.7.:** (a) Normalised stress distribution and (b) craze contour in terms of normalised fibril length along ligament at four loading stages.

The influence of the fibril creep viscosity $\eta_c$ on the ligament stress and the evolution of the craze and crack length is studied over a range of four orders of magnitude of the dimensionless parameter $\tau_c/T_0$ in Figure 4.8. Higher $\tau_c/T_0$ values enhance fibril stiffness,

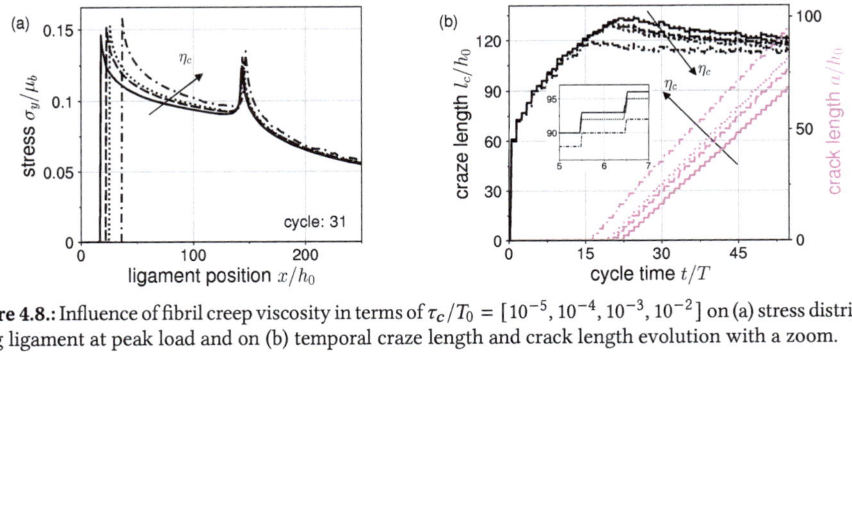


**Figure 4.8.:** Influence of fibril creep viscosity in terms of $\tau_c/T_0 = [10^{-5}, 10^{-4}, 10^{-3}, 10^{-2}]$ on (a) stress distribution along ligament at peak load and on (b) temporal craze length and crack length evolution with a zoom.

resulting in higher stress, which is most pronounced at the crack tip with the longest craze fibrils $\xi_0$ (cf. Figure 4.8(a)). In all cases, craze initiation occurs at the notch tip at 40% of the peak load during the first loading cycle. Thereafter, the craze length $l_c$ monotonically grows over multiple cycles prior to failure, representing cyclic damage accumulation (cf. Figure 4.8(b)). The inset in Figure 4.8(b) illustrates that craze growth is interrupted by arrest phases during unloading. Crack initiation also takes place at the notch tip and is delayed from the 17$^{\mathrm{th}}$ to the 23$^{\mathrm{rd}}$ loading cycle as $\tau_c/T_0$ decreases. The correlation between $\sigma_y$ and $\tau_c/T_0$ results in a faster crack propagation after initiation, whereas the craze length $l_c$ shortens with increasing $\tau_c/T_0$.

The influence of the loading magnitude is analysed by increasing the overall deformation by 20% in Figure 4.9, allowing for comparisons with experimental findings discussed below. The craze contour at peak load in cycles with approximately equal crack extensions is shown in Figure 4.9(a). The evolution of the craze and crack length during the cyclic loading history is depicted in Figure 4.9(b). The craze width at the crack tip $\xi(x = a)$, the craze length $l_c$ and the crack growth rate all increase with $\Delta u_y$. However, the most notable difference is the change in the evolution of $l_c$, which exhibits a more pronounced variation for the higher loading amplitude.

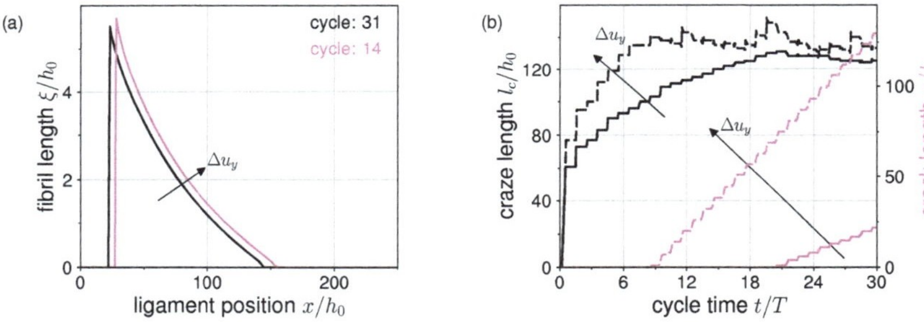


**Figure 4.9.:** Influence of load increase by 20% on (a) craze contour and (b) craze and crack length evolution.

### 4.3.3. Discussion

The simulation results, including the parameter choice, are assessed by drawing comparisons to experiments and previous numerical analyses from the literature. The double stress peak at the crack tip and the craze tip is in accordance with previous results for various brittle glassy polymers, e.g. in [15, 71, 99, 103], who calculated the normal stress distribution along a craze based on experimentally measured craze opening contours and the assumption of a linear elastic bulk response. A non-proportional stress redistribution along a craze during a loading-unloading cycle similar to Figure 4.7(a) is also reported in [71]. In addition, the stress profile at unloading computed in [15] shows a compressive stress at the crack tip and a slight tensile stress at the craze tip, which corresponds to the possible effect of fibril jamming at the unloaded crack tip and agrees well with our result (orange line) in Figure 4.7(a). A further quantity which has been studied extensively is the craze profile, from which characteristic ratios of craze length vs. craze opening displacement $l_c/(2\xi_{max})$ can be derived. A representative range for glassy polymers is $l_c/(2\xi_{max}) = 10 - 20$ (cf., e.g., [23]), while the results presented here yield $l_c/(2\xi_{max}) \approx 11.3$, which is well within the experimental range. We remark, however, that this ratio is influenced by the choice of craze parameters, for instance, $\lambda_c^* = 3$ yields $l_c/(2\xi_{max}) \approx 14$. Finally, increasing the loading amplitude results in an increase in craze length $l_c$ and craze opening displacement at the crack tip $\xi(x = a)$ due to the modelled viscosities, which also aligns with experimentally observed trends [50].

# 4.4. Interaction between crazing and shear yielding in ductile glassy polymers

## 4.4.1. Modelling aspects

In the following, we study the interaction of crazing and shear yielding under cyclic mode I crack growth as it may occur in more ductile glassy polymers. This extends the work by Van der Giessen and co-workers who considered monotonic loading conditions [26, 27]. The objective far field stress outlined in Subsection 4.3.1 is obtained by the maximum deformation $\varepsilon_{y,max} = 3.3 \cdot 10^{-3}$, resulting in an average far field stress of $\sigma_y \approx 10$MPa in the simulations. Additionally, following the cyclic loading conditions outlined by Takemori [94], the load ratio is set to $R_u = 0.1$. This load ratio enables to exploit mode I symmetry, as it mitigates the effect of crack closure due to compression at the crack tip resulting from the crack tip plastic zone (cf. [73]). The viscoplastic behaviour of the bulk material is described by a modified version [37] of the standard glassy polymer model by Boyce [17] (see Appendix 4.B for details). The bulk parameters, which are representative for PC, are outlined in Table 4.B.1. The analysed craze materials in this section are given in Table 4.1 and the drawing and fibril creep viscosities are taken as $\tau_d/T_0 = \tau_c/T_0 = 10^{-4}$. This choice, in conjunction with the nonlinear Eyring-type viscosity of the bulk model (cf. Appendix 4.B), guarantees that in the following numerical studies, focusing only on a generic glassy polymer, all inelastic mechanisms (fibril drawing and creep as well as bulk shear yielding) are active.

## 4.4.2. Numerical results

At first, craze material 2 (cf. Table 4.1) is investigated. The field output for the accumulated plastic strain $\gamma^p$, i.e. the plastic zone, is displayed in Figure 4.10 for three instants in the first cycle: just prior to craze initiation (Figure 4.10(a)), after craze initiation (Figure 4.10(b)) and at load maximum (Figure 4.10(c)). All instants fall within the first (monotonic) loading stage, allowing for comparisons with previous work [26]. The grey area along the ligament shows the location of the un-crazed elements, while the black area indicates craze elements after initiation, i.e. $\xi_0 > 0$, and therefore the craze length $l_c$. The following is notable: Craze initiation, occurring at $\varepsilon_y \approx 0.4\varepsilon_{y,max}$, is preceded by shear yielding (cf. Figure 4.10(a) and (b)). Although the overall plastic zone is still small in size and magnitude (cf. Figure 4.10(b) and (c)), craze initiation takes place at an approximately 20% higher load compared to the elastic bulk response (cf. Subsection 4.3.2). Additionally, the craze initiates ahead of the notch root at the tip of the plastic zone where the local stress $\sigma_y$ attains a maximum (cf. Figure 4.10(b)). Upon initiation, the craze grows in both directions, towards the notch tip and along the ligament (cf. Figure 4.10(c)), while the plastic zone develops the characteristic plane strain pattern observed in experiments, e.g. [45], and previous simulations, e.g. [26, 27]. Craze initiation ahead of a plane strain notch is consistent with experimental results in PC [44, 45], where the craze initiated at the intersecting shear bands. Furthermore, it aligns with simulation results reported in [26, 27], where the location of craze initiation shifted from the notch tip into the bulk material with increasing craze initiation stress.

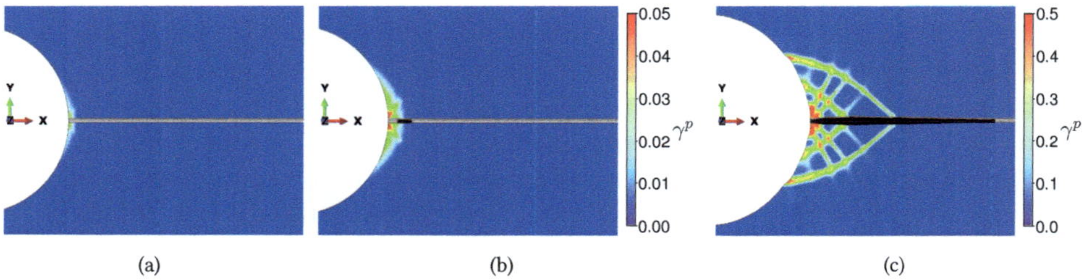


**Figure 4.10.**: Accumulated plastic strain $\gamma^p$ in first cycle during loading (a) at instant prior to craze initiation, (b) at instant after craze initiation and (c) at load maximum. Black zone along ligament indicates craze. Data for craze material 2 in Table 4.1.

The response of craze material 2 in the course of continuous cyclic loading is analysed using snapshots of the plastic zone and craze evolution. For this, Figure 4.11 displays contour plots of $\gamma^p$ at three instants: after crack initiation in the 4$^{\text{th}}$ cycle (Figure 4.11(a)) and at peak load in the 6$^{\text{th}}$ (Figure 4.11(b)) and the 8$^{\text{th}}$ cycle (Figure 4.11(c)). To enhance visualisation, the colour bar is capped at $\gamma^p = 1.5$, with the maximum value also indicated. After craze initiation, fibril damage accumulates over three cycles through fibril drawing, leading to craze breakdown at the outermost intersecting shear band in the 4$^{\text{th}}$ cycle (cf. Figure 4.11(a)). To account for the Bauschinger effect in glassy polymers, the shear yielding model (cf. Appendix 4.B) incorporates kinematic hardening, resulting in a continuous increase in the $\gamma^p$ magnitude with each loading cycle. Despite the quantitative increase, the plastic zone remains qualitatively stationary after the 1$^{\text{st}}$ cycle (cf. Figure 4.10(c) and Figure 4.11(a)). With further cyclic loading, the crack propagates in both directions and connects with the notch root (cf. Figure 4.11(c)). Right at the load maximum in the 6$^{\text{th}}$ cycle, thin shear bands form at the right-hand side crack tip (cf. Figure 4.11(b)). A weaker shear band to its left forms during unloading in the 5$^{\text{th}}$ cycle. Both shear bands arch forward in the direction of the ligament. Each subsequent load cycle is accompanied by the formation of two additional shear bands originating from the propagating crack tip (cf. Figure 4.11(c)). The first band forms at the load maximum and a second, less pronounced band during unloading, leading to the fracture pattern observed in Figure 4.11(c).

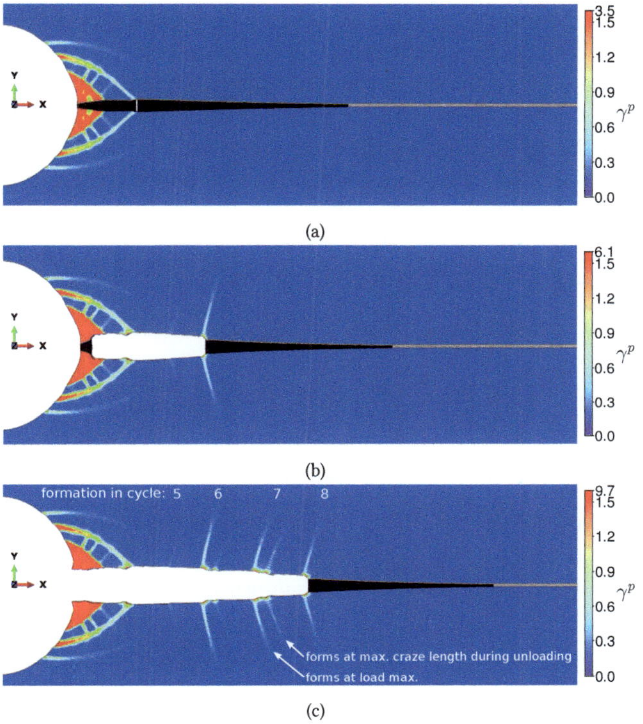


**Figure 4.11.:** Accumulated plastic strain $\gamma^p$ (a) at crack initiation in cycle 4 and at load maximum in (b) cycle 6 and (c) cycle 8. Black zone along ligament indicates craze. Data for craze material 2 in Table 4.1.

The cyclic response is further investigated by examining the evolution of the craze length $l_c$ and crack length $a$ over eight consecutive loading cycles shown in Figure 4.12(a) and with a detail view of the 7$^{\text{th}}$ cycle shown in Figure 4.12(b). Figure 4.12(a) is supplemented with the normalised loading programme (green line) and three vertical grey dotted lines indicating the start of craze growth during loading, the load peak and the end of craze growth during unloading in the 2$^{\text{nd}}$ cycle. Similar to the case of purely elastic bulk material (cf. Figure 4.8(b) and Figure 4.9(b) in Subsection 4.3.2), the craze length grows continuously until crack initiation. The vertical dotted lines indicate that growth of $l_c$ is confined to a range around load maximum prior to crack initiation. However, since the lines are not equidistantly spaced, $l_c$ increases asymmetrically with respect to $\varepsilon_{y,max}$. The craze length rapidly drops at crack initiation, primarily due to the computation methodology of $l_c$ outlined in Subsection 4.3.1. The further decline of $l_c$ in the 5$^{\text{th}}$ cycle is indeed associated with a shrinking craze zone of the right-hand side craze in Figure 4.11(a). After crack initiation, the crack advances continuously with each cycle. The initial crack growth during the remainder of the 4th cycle is small, as initiation occurs during unloading. Subsequently, the crack propagates in both directions (cf. Figure 4.11(b)) until it reaches the notch tip in the 7th cycle. The transition from two propagating crack tips to one combined with an initially higher crack velocity in the direction of the notch results in a large crack growth increment $\Delta a$ in the 5th cycle, which thereafter declines. Throughout one cycle, $l_c$ initially decreases while crack propagation commences (cf. Figure 4.12(b)). That is, the leading edge of the craze remains stationary while the trailing edge at the crack tip advances. Coinciding with the formation of the first shear band at load maximum (cf. Figure 4.11(c)), the leading edge of the craze accelerates, while the crack speed slightly declines (cf. Figure 4.12(b)). The formation of the second shear band in Figure 4.11(c) occurs right at the $l_c$ peak during unloading.

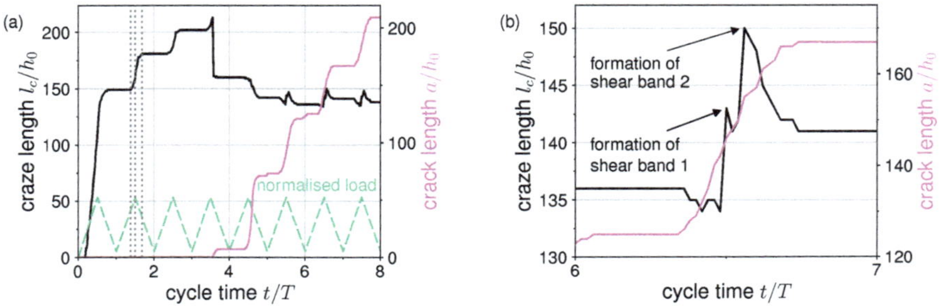


**Figure 4.12.:** Craze length (black) and crack length (magenta) evolution for (a) 8 loading cycles with normalised load (green) and (b) detailed view of 7th cycle. Data for craze material 2 in Table 4.1.

To elucidate the role of the crack tip plastic zone on the crack and craze advancement, craze material 1 (cf. Table 4.1) with $\mu_b/\mu_f = 1$ is investigated in the following. With the bulk parameters remaining the same, the change in material corresponds to a higher fibril stiffness. Likewise to above, snapshots of the plastic zone and craze evolution at three instance – featuring crack initiation in the 2nd cycle and load maxima in the 3rd and 6th cycle – are shown in Figure 4.13. At crack initiation, the plastic zone resembles that of the previous material (cf. Figure 4.11(a) and Figure 4.13(a)). Moreover, crack initiation essentially occurs at the same location. However, the higher fibril stiffness results in higher craze stress (cf. Figure 4.4), leading to accelerated fibril drawing and, consequently, the earlier crack initiation in the 2nd cycle. Similarly, this leads to faster crack propagation and elevated plastic activity, resulting in a much more pronounced plastic zone depicted in Figure 4.13(c).

For material 1, the quantitative evolution of the fracture process over six consecutive loading cycles is shown in Figure 4.14, displaying the craze length $l_c$ (black line) and crack length $a$ (magenta line) with a detail view of the 5th cycle. Additionally, Figure 4.14(b) includes three vertical grey dotted lines indicating the load maximum and two equidistant instants, where the left line marks the instant before the formation of a plastic zone at the current crack tip. The following traits are qualitatively similar to those of material 2: First, craze initiation is succeeded by a monotonic increase in $l_c$ until crack initiation.

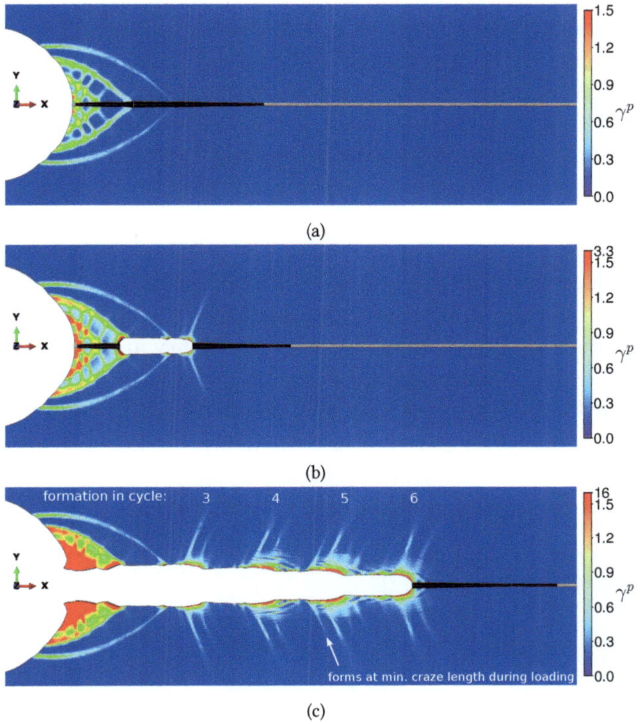


**Figure 4.13.:** Accumulated plastic strain $\gamma^p$ (a) at crack initiation in cycle 2 and at load maximum in (b) cycle 3 and (b) cycle 6. Black zone along ligament indicates craze. Data for craze material 1 in Table 4.1.

Second, the crack propagates continuously in each cycle, with the largest crack increment occurring in the cycle immediately after initiation. Finally, the trailing and leading edges of the craze advance at different velocities throughout a cycle, resulting in the non-monotonic variation of $l_c$.

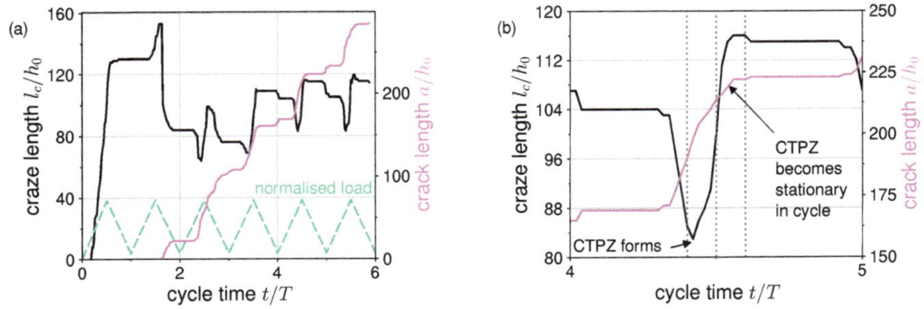


**Figure 4.14.:** Craze length (black) and crack length (magenta) evolution for (a) 6 loading cycles with normalised load (green) and (b) detailed view of 5th cycle. Data for craze material 1 in Table 4.1.

The role of the crack tip plastic zone can be deduced from Figure 4.14(b) in combination with Figure 4.15, which shows snapshots of the plastic zone for the three instants marked by the vertical lines in Figure 4.14(b). Crack growth starts at $\varepsilon_y = 0.64\varepsilon_{y,max}$ and is primarily confined to the loading stage (i.e. $t/T < 4.5$). The crack propagates initially into an elastic medium (cf. Figure 4.15(a)) and without energy dissipation of the bulk, the crack growth is comparatively fast. However, it significantly slows down once a crack tip plastic zone (CTPZ) develops in form of shear bands during loading at $\varepsilon_y = 0.84\varepsilon_{y,max}$, coinciding with the minimum craze length in Figure 4.14(b). The plastic zone continues to develop (cf. Figure 4.15(b)), while crack growth slows down. Simultaneously, $l_c$ rapidly increases from $83h_0$ to $116h_0$, corresponding to a rise of approximately 40% in craze length. Just preceding crack arrest, the crack tip plastic zone becomes stationary and maintains the shape in Figure 4.15(c). The crack arrests during unloading at a load level of $\varepsilon_y = 0.76\varepsilon_{y,max}$, which is nearly 20% higher than the load that initiated crack growth in that cycle. The pronounced plastic zone prior to $\varepsilon_{y,max}$ and the delayed crack growth behaviour is in contrast to material 2 (cf. Figure 4.11 and Figure 4.12(b)). Hence, we conclude that the crack tip plastic zone impedes significantly the crack advance.

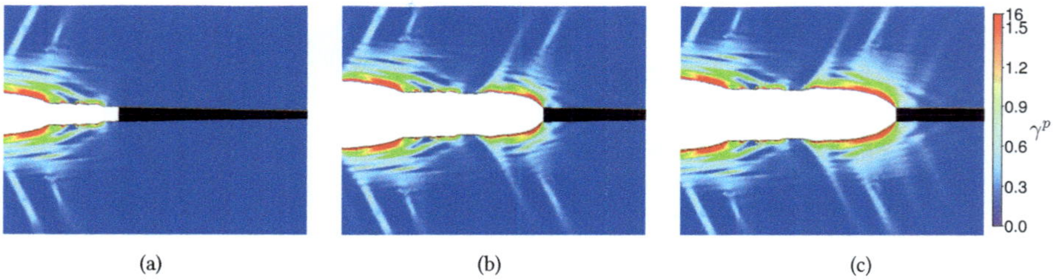


**Figure 4.15.:** Detail of crack tip plastic zone in 5$^{th}$ cycle (a) instant just prior to crack tip plastic zone formation at $t/T = 4.4$, (b) at peak load $t/T = 4.5$ and (c) during unloading at $t/T = 4.6$. Vertical lines in Figure 4.14(b) indicate instants. Black zone along ligament indicates craze. Data for craze material 1 in Table 4.1.

To investigate the driving mechanism behind the observed delayed crack propagation, the craze contour is shown in Figure 4.16(a) for material 2 and in Figure 4.16(b) for material 1 at equidistant time intervals of $\pm 0.1\Delta t/T$ around peak loading at the indicated load cycle. Note that for material 1 (Figure 4.16(b)), these time instants correspond to the snapshots in Figure 4.15 and the vertical lines in Figure 4.14(b). The following trends apply to both materials but are much more pronounced for material 1 (Figure 4.16(b)): Prior to crack tip plasticity (black lines), the craze width monotonically declines from the crack tip. Once the crack tip plastic zone forms, the craze contour exhibits a locally confined plateau at the crack tip (magenta and green lines). This plateau corresponds to a plasticity-induced local unloading of the craze fibrils, which impedes crack propagation. Further fibril growth is necessary before additional crack advancement can occur, which is enabled by the increase in craze length $l_c$ seen in Figure 4.14(b). The plateau is most pronounced for the largest crack tip plastic zone (green line in Figure 4.16(b)), where crack propagation is most impeded. Concluding, this analysis suggests that the delayed crack propagation results from the unloading of the crack tip due to the crack tip plastic zone. Subsequently,

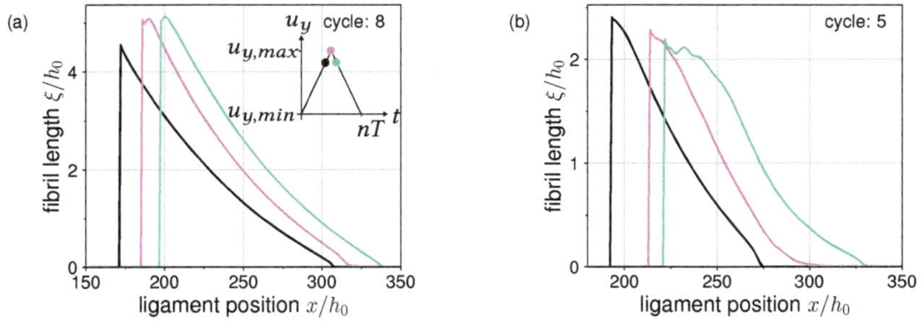


**Figure 4.16.:** Craze contour at equidistant time intervals around peak loading ($\pm 0.1\Delta t/T$) of indicated cycle for (a) craze material 2 and (b) craze material 1 for which the vertical lines in Figure 4.14(b) indicate instants.

$l_c$ must first grow (cf. Figure 4.14(b)) to reach a critical craze width, which then permits further crack advance.

Reducing the limit stretch $\lambda_L$ in the craze model so it aligns with the MD results (cf. material 3 in Table 4.1), yields similar results to material 2. For conciseness, results are therefore only presented and briefly discussed in Appendix 4.C.

### 4.4.3. Discussion

The name-giving shape of the *epsilon crack tip plastic zone* where shear bands and a craze emanate from the crack tip was reported by Mills and Walker [63] and extensively studied by Takemori, e.g. [92, 94], in the context of *discontinuous crack growth* under cyclic (fatigue) loading. The formation of shear bands increases the load bearing capacity, which Takemori [94] attributed to a shielding of the craze by reducing the hydrostatic stress. The simulation results (cf. Figure 4.11 and Figure 4.13) exhibit qualitative similarity to those presented in Fig. 2 and Fig. 15 of [94][2]. In both cases, shear bands initiate at the crack tip and arch towards the craze. Moreover, the craze length varies in the course of crack propagation which is a feature also observed in discontinuous crack growth. However, unlike the experimental observations, our simulations show *normal crack growth*, i.e. the crack propagates in each cycle and additionally the craze length varies throughout one cycle. Nonetheless, a noteworthy similarity is observed in the *delayed crack propagation* once a pronounced plastic zone forms. The mechanisms driving this crack deceleration and potential arrest are attributed to local crack tip plasticity-induced unloading. This provides a physical explanation for the delaying crack propagation, which may be extended to discontinuous crack growth.

For materials with less crack tip plasticity, such as material 2, the simulation results align well with the experimental picture obtained from measuring fatigue striations as discussed in the review [23]. In these cases, the crack advances predominantly near the

[2] See also Figure 1.2 in Section 1.1

load maximum. Moreover, the crack growth precedes craze growth, so that the crack initially propagates into the stationary craze (cf. Figure 4.12(b)). Once crack and craze growth arrests, the craze length returns approximately to its initial length at the beginning of the cycle (cf. Figure 4.12(a)). Yet, the simulations provide additional insights, showing that the craze length rapidly increases and ceases to grow *before* the crack arrests.

## 4.5. Concluding remarks

Craze and crack propagation in glassy polymers under cyclic mode I loading were investigated numerically by employing a recently developed continuum-micromechanical model for crazing. A particular challenge was the parameter calibration, as craze and bulk parameters need to be chosen consistently to yield a generic, but representative glassy polymer response. To address this, the bulk model parameters were based on a glassy polymer calibration from the literature. Subsequently, the craze parameters were normalised with the bulk parameters accordingly. Parameter scopes of the craze model, which cannot be obtained from the experimental literature, were identified using molecular dynamics simulations of the cyclic craze response.

Two types of craze and crack growth simulations were conducted, namely in brittle and in ductile glassy polymers. The craze model reproduces important craze characteristics from the experimental literature for brittle glassy polymers. This includes measurements such as the ratio of craze length to craze width, the craze contour and experimental trends, for instance, the increased crack opening displacement and craze width with higher load amplitudes due to the modelled viscosities. Moreover, similar to experimental findings on *normal fatigue crack growth*, both, crack and craze, grew near load maximum and arrested during unloading. The simulation results also aligned with previous experimental and theoretical results, indicating a double stress peak at the crack and craze tips and compression at the crack tip during unloading. Additionally, building upon work by Van der Giessen and co-workers [26], the competition of crazing and shear yielding under cyclic loading for ductile glassy polymers was analysed. The presence of a plastic zone was found to increase the load for craze initiation and, in accordance with experiments [45] and previous simulations [26], craze and crack initiation occurred ahead of the notch root. The plastic zone exhibited similarity with the epsilon crack tip plastic zone observed in cyclic (fatigue) loading. The model replicated the initiation of shear bands at the current crack tip, which grew and arched towards the craze. A salient finding of this study was the delayed crack propagation due to crack tip plasticity. This delay was attributed to plasticity-induced unloading of the craze adjacent to the crack tip, which required additional craze growth prior to further crack advance. The size of the unloaded area, influencing the delay, correlated to the amount of crack tip plasticity.

The study faces some limitations: First, the results were obtained using a generic glassy polymer and hence, limiting the results to qualitative assessment. Second, the simulations do not yet capture the so-called *discontinuous* or *retarded* crack growth, where the crack advances by jumps after remaining stationary for hundreds of loading cycles. It is hypothesised that thermal fatigue of the craze fibrils in close vicinity to the crack tip might influence this behaviour (cf., e.g., [23]). We, on the other hand, considered only isothermal conditions. However, insights from experiments and molecular dynamics suggest a hysteretic craze response. The associated dissipation could be a source of heating and hence, adiabatic or coupled thermo-mechanical simulations could offer valuable insight. Yet, this approach presents significant challenges as it demands accurately determining the mechanical response of crazes at elevated temperatures.

Concluding, an interesting direction for future work appears to be experimental investigations of the cyclic craze response. An emphasis on the cyclic uniaxial craze behaviour across various strain rates and temperatures will be of great value. This data can contribute to (in-)validate the molecular dynamic results, which would help with additional physically motivated model refinements.

# Appendix to Chapter 4

## 4.A. Parameter calibration via molecular dynamics simulations of cyclic crazing

The molecular dynamics (MD) based calibration builds upon recent work [55] where a generic bead-spring model was used to investigate the cyclic craze response in glassy polymers subjected to uniaxial deformation. The focus here is solely on utilizing the results to estimate the material parameters. Since the MD model is standard and has been extensively used in previous craze studies (e.g. [32, 76]), the reader is referred to [55] for details on model and simulation. In the following, the overall stretch and the axial stress of the MD system is denoted by $\lambda$ and $\sigma$, respectively. They correspond to $\bar{\lambda}_1$ and $\bar{\sigma}_1$ of the craze model in Section 4.2. The loading programme, the stress response and a snapshot of the craze at peak deformation is shown in Figure 4.A.1. The initially isotropic glass is uniaxially deformed to a maximum stretch $\lambda_{max} = 10$ (grey dotted line in Figure 4.A.1(a) and in favour of visibility only partially shown in (b)), which leads to a full conversion of bulk material to fibrillated craze matter (cf. snapshot in Figure 4.A.1). Subsequently, the craze is subjected to a cyclic loading programme with varying unloading magnitudes $\lambda_{min} = [1, 2, 3, 5, 8]$, followed by reloading to $\lambda_{max}$. The stress response of the five independent simulations is shown in Figure 4.A.1(b) exhibiting a hysteresis and a reloading behaviour which depends on $\lambda_{min}$. Furthermore, it is noteworthy that the craze exhibits a pronounced compressive stress (cf. black dashed line Figure 4.A.1(b))

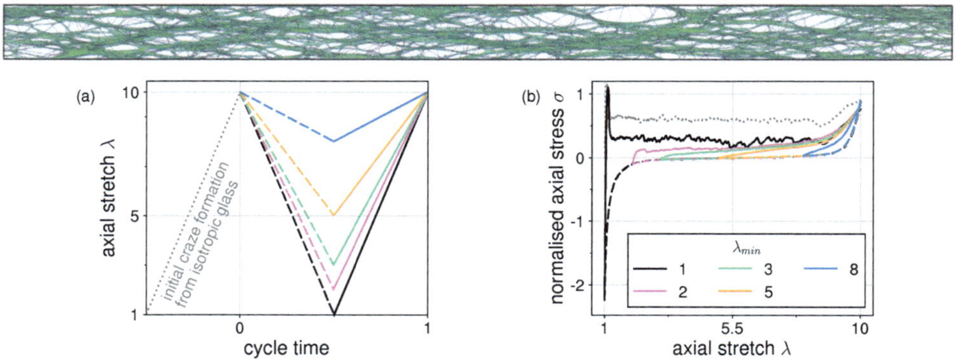


**Figure 4.A.1.:** Snapshot from MD simulation for craze at $\lambda_{max}$ = 10 (top) and (a) uniaxial deformation controlled cyclic loading programme with (b) stress response of sole fibrillated craze matter for five unloading magnitudes $\lambda_{min}$.

prior to reaching the macroscopic undeformed state $\lambda = 1$. This stress arises from the intermolecular resistance (cf. [55]) and is macroscopically interpreted as fibril jamming.[3]

Additional useful information from the MD simulations is the interaction between bulk and craze material subjected to uniaxial cyclic loading. Those bulk-craze systems are created by reducing $\lambda_{max}$. Scaling the deformation in terms of the engineering strain of each system with its value at $\lambda_{max}$ defines the scaled engineering strain $\varepsilon_E = (\lambda - 1)/(\lambda_{max} - 1)$. The thus rescaled stress-strain response is depicted in Figure 4.A.2 for sole craze matter ($\lambda_{max}$=10) and two additional bulk-craze systems. To provide a notion of the bulk-craze composition for $\lambda_{max} = 5$ and $\lambda_{max} = 2$, snapshots for the two systems at their respective $\lambda_{max}$ are also displayed in Figure 4.A.2. An interesting observation is the quasi-collapse of the stress-strain curves, which implies a scaling of the craze response with $\lambda_{max}$. This, on the other hand, is related to the maximum fibril length as discussed in detail in [55] and which aligns well with the micromechanical framework of the continuum model.

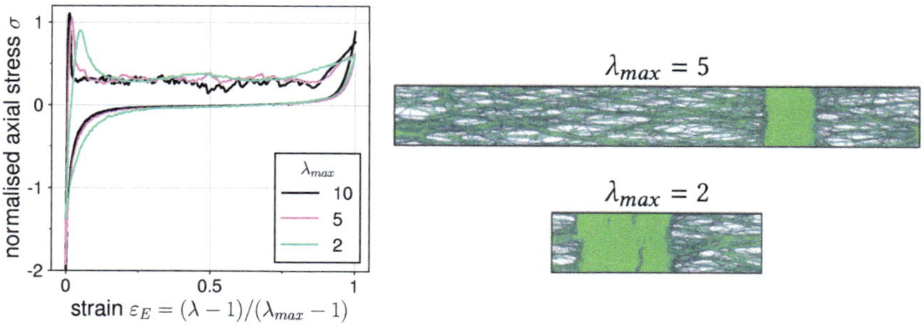


**Figure 4.A.2.:** Quasi-collapse of hysteresis for several bulk-craze compositions $\lambda_{max}$ when scaled to $\varepsilon_E$ (left) and snapshots for $\lambda_{max} = [2, 5]$ (right).

[3] For a detailed analysis of the driving mechanisms leading to this mechanical response, the reader is referred to [55].

Insights from Figure 4.A.1 and Figure 4.A.2 are used in the following to identify the parameters of the elastic behaviour of the fibrillated craze matter ($\mu_b/\mu_f$, $\mu_{f,NH}/\mu_f$ and $\lambda_L$), the deformation at which fibril jamming occurs during unloading (parameter $\beta_c$ in (4.13)) and the fraction upon which mature fibrils exist (parameter $\alpha$ in Figure 4.2).

## 4.A.1. Elastic craze fibril parameters: $\mu_b/\mu_f$, $\mu_{f,NH}/\mu_f$ and $\lambda_L$

Although $\lambda_c^*$ is taken from experiments, $\lambda_c^*$ from the MD simulation is needed to determine the elastic craze fibril parameters $\mu_b/\mu_f$, $\mu_{f,NH}/\mu_f$ and $\lambda_L$. Recall that $\lambda_c$ describes the density-ratio in the unloaded configuration (cf. (4.2)). Hence, the bulk density $\rho_b$ is taken as that of the initially undeformed isotropic glass. In contrast, the craze density $\rho_c$ is computed at the instant $\sigma = 0$ during unloading (dashed lines in Figure 4.A.1(a)), which is assumed to approximately coincide with the unloaded configuration, yielding a MD based value of $\lambda_c^* \approx 6.5$. This high value is an inherent issue with the MD model, which poses challenges for transferring information from the MD model to the continuum model. This issue is addressed by normalising the MD based parameters and investigating a range of parameters associated with the elastic part of the fibril model in the continuum model (cf. Table 4.1).

With $\lambda_c^* \approx 6.5$, a least square fitting of the elastic craze fibril model is employed for the parameter calibration. Using only the elastic part of the craze fibril behaviour is based on the following considerations: To separate the bulk and craze response and to avoid potentially transient behaviour while bulk and craze coexist, e.g. due to fibril drawing, the cyclic MD results for sole fibrillated craze matter shown in Figure 4.A.1 are used as fitting data. This corresponds to the viscoelastic fibril deformation behaviour of the continuum model. Furthermore, the hysteresis in Figure 4.A.1(b) exhibits negligible rate dependencies as discussed in [55]. Combined with the inherently high deformation rates in MD simulations, we concluded that the dashpot in the craze fibril model is most appropriately taken as stiff during calibration with the MD data. This reduces the viscoelastic fibril response to the purely elastic fibril response of two springs in parallel.

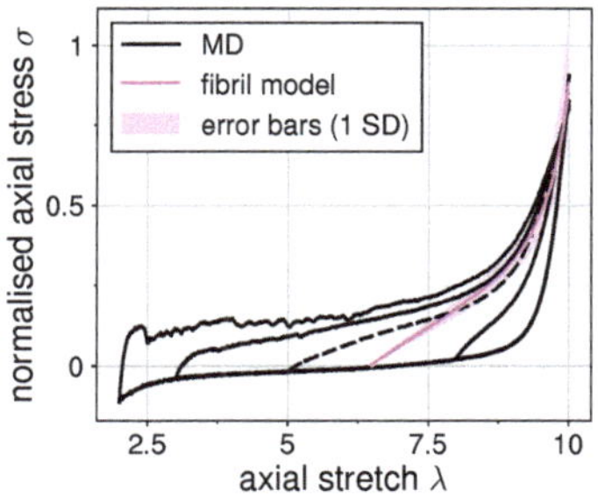


**Figure 4.A.3.:** Least square fit of elastic part of craze fibril model based on MD result for $\lambda_{min} = 5$ (dashed line).

For simplicity, the continuum model is fitted to the results for $\lambda_{min} = 5$, which comprises two advantages: Firstly, fibrils are assumed to be loose hanging for deformations smaller than $\lambda_c = 6.5$, which is sufficiently close to $\lambda_{min} = 5$. Secondly, $\lambda_{min} = 5$ avoids the complicated case of pore space closure for $\lambda \leq 2$, altering significantly the structural craze response (cf. [55]). The result of the fit is shown in magenta in Figure 4.A.3 with one standard deviation error bars. The black lines represent the MD result already shown in Figure 4.A.1(b), but with the exception that the case $\lambda_{min} = 1$ is dropped to enhance visibility. The dashed line is the objective data used in the fitting. To obtain the ratio of bulk and fibril stiffness, the bulk stiffness is calculated as the secant stiffness of the glass prior to cavitation. This leads to $\mu_b/\mu_f = 10.7$, $\mu_{f,NH}/\mu_f = 0.48$ and $\lambda_L = 1.13$.

## 4.A.2. Fibril jamming: parameter $\beta_c$

Fibril jamming is observed in the MD simulations for $\lambda_{min} \leq 2$ in Figure 4.A.1(b) and more generally for different craze-bulk compositions below $\varepsilon_E < 0.2$ in Figure 4.A.2. To make use of (4.13), $\lambda_{max}$ needs to be related to $\xi_0$, which is in the MD simulations the maximum attained craze length throughout deformation. Likewise to above, it is assumed that the configuration at $\sigma = 0$ during unloading closely resembles the unloaded configuration and hence, $\xi_0$ and $\chi_0$ are calculated at $\sigma = 0$. Figure 4.A.2 is then rescaled with $\hat{\varepsilon}_1$ from (4.12), which defines the scaled Hencky strain $\varepsilon_H = \ln\lambda/\ln\hat{\lambda}_1$, leading to Figure 4.A.4. This relates the deformation to the fibril length $\xi_0$ and as additional benefit, it improves the collapse for $\varepsilon_H < 0.25$, which is the region of interest here.

As described in Subsection 4.2.1, fibril jamming is considered via the elastic bulk response. Therefore, it is approximated in a simplified linear manner indicated by the orange lines in Figure 4.A.4. The lines represent the secants connecting $\varepsilon_H = 0$ and the normalised stress $\sigma = -0.5$. Latter comprises a subjective component and was selected here due to its relevant stress magnitude with respect, for instance, to the drawing stress plateau. This results in $\beta_c \approx 0.15$, which is indicated by the black vertical marker in Figure 4.A.4.

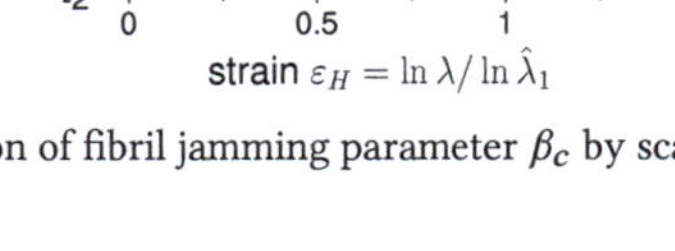


**Figure 4.A.4.:** Estimation of fibril jamming parameter $\beta_c$ by scaling deformation to $\varepsilon_H$.

### 4.A.3. Morphology change: parameter $\alpha$

To identify $\alpha$ depicted in Figure 4.2, the instant upon which mature fibrils exist during craze formation from the isotropic glass, i.e. where the density reaches $\rho_c^* = \rho_b/\lambda_c^*$, is identified. For this, the evolution of the density profile (colour coding) in axial direction (abscissa) as function of the deformation (ordinate) is shown in Figure 4.A.5. The initial normalised bulk density ($\rho_b \approx 1$) rapidly drops during cavitation. As assumed in the continuum model, the MD results show that the creation of mature fibrils is a continuous process requiring further overall deformation upon cavitation. The instant $\rho = \rho_c^*$ is indicated by the black horizontal line. This configuration is then used to calculate $\alpha$ via the mass balance: $\alpha = 1 - \chi/h_0$, which assumes $\chi_0 \approx \chi$. For a normalised mature craze density of $\rho_c^* = 0.18$, this yields $\alpha \approx 0.07$.

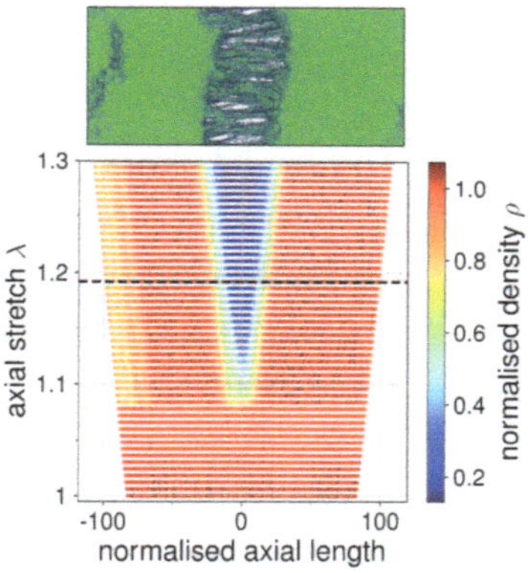


**Figure 4.A.5.:** Normalised density distribution (colour coding) along axial simulation box length (abscissa) evolving with deformation (ordinate). MD snapshot at $\lambda = 1.3$ indicates local density distribution and serves to facilitate plot interpretation. Black dashed horizontal line marks instant upon which mature fibrils have formed.

Similar to experiments, a general difficulty arising from MD simulations is its statistical nature. This results in a certain subjectivity when selecting thresholds such as the craze density $\rho_c^*$. However, in this case, the influence of $\rho_c^*$ is relatively insensitive as, e.g., $\rho_c^* = 0.14$ leads to $\alpha = 0.08$.

## 4.B. Constitutive model for shear yielding

Building upon the early work by Boyce et al. [17], there are well established constitutive models describing finite strain shear yielding in glassy polymers, e.g. [4, 9, 37, 39, 106]. We use the version by Hempel [37], for which the rheological model and its components are shown in Figure 4.B.6 and which is briefly summarised in the following. The rheological

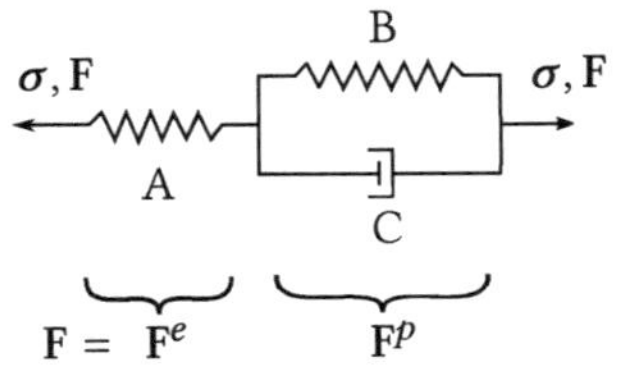


Constitutive model: (material parameters)
A: neo-Hooke hyperelasticity: $(\mu_b, \nu_b)$
B: 8-chain hyperelasticity: $(C_R, \lambda_L)$
C: Eyring-type viscous flow: $(\dot{\gamma}_0^p, A, s_0, s_s, h, \alpha_p)$

**Figure 4.B.6.:** Rheological model for bulk material.

model leads to the standard multiplicative decomposition of the deformation gradient (cf. [4])

$$\mathbf{F} = \mathbf{F}^e \mathbf{F}^p \tag{4.26}$$

into an elastic $\mathbf{F}^e$ and plastic contribution $\mathbf{F}^p$, yielding the elastic right $\hat{\mathbf{C}}^e = \mathbf{F}^{e\,\mathrm{T}}\mathbf{F}^e$ and inelastic left Cauchy-Green tensor $\hat{\mathbf{b}}^p = \mathbf{F}^p \mathbf{F}^{p\,\mathrm{T}}$ in the intermediate (i.e. relaxed) configuration $\hat{\Omega}$. The velocity gradient $\mathbf{l}$ is then additively split

$$\mathbf{l} = \dot{\mathbf{F}}\mathbf{F}^{-1} = \mathbf{l}^e + \mathbf{F}^e\,\hat{\mathbf{l}}^p\mathbf{F}^{e\,-1}, \tag{4.27}$$

where $\mathbf{l}^e = \dot{\mathbf{F}}^e\dot{\mathbf{F}}^{e-1}$ is the elastic velocity gradient in the current configuration $\Omega$ and $\hat{\mathbf{l}}^p$ is the plastic component in $\hat{\Omega}$, which can be further decomposed in the symmetric rate of deformation tensor $\hat{\mathbf{d}}^p$ and the skew-symmetric inelastic spin tensor $\hat{\mathbf{w}}^p$:

$$\hat{\mathbf{l}}^p = \operatorname{sym}\hat{\mathbf{l}}^p + \operatorname{skw}\hat{\mathbf{l}}^p = \hat{\mathbf{d}}^p + \hat{\mathbf{w}}^p. \tag{4.28}$$

The constitutive description of the model is placed in the intermediate configuration $\hat{\Omega}$, for which the isotropic neo-Hooke hyperelasticity is given by the Mandel stress

$$\hat{\Sigma} = \mu_b(\hat{\mathbf{C}}^e - \mathbf{1}) + \bar{\lambda}_b\,(J^e - 1)\,J^e\mathbf{1}, \tag{4.29}$$

where $J^e = \det \mathbf{F}^e$ and which is symmetric due to the assumption that the elastic free energy contribution $\psi^e$ and inelastic free energy contribution $\psi^p$ are isotropic functions of $\hat{\mathbf{C}}^e$ and $\hat{\mathbf{b}}^p$, respectively. The model features the two Lamé parameters $\bar{\lambda}_b$ and the shear modulus $\mu_b$. Note, as the difference is negligible for small elastic deformations, we omit correcting the elastic bulk constants due to the different elasticity models in (4.17) and (4.29) for the craze and bulk model, respectively.

The resistance against plastic molecular network deformation is described by the eight-chain model [8]

$$\hat{\boldsymbol{\tau}}^b = \frac{C_R\lambda_L}{3\lambda_C}\,\mathcal{L}^{-1}(\lambda_C/\lambda_L)\hat{\mathbf{b}}^{p\prime}, \tag{4.30}$$

where $\hat{\mathbf{b}}^{p\prime}$ is the deviatoric part of $\hat{\mathbf{b}}^p$ and $\lambda_C$ is the the mean chain stretch reading

$$\lambda_C = \left(\frac{\operatorname{tr}\hat{\mathbf{b}}^p}{3}\right)^{1/2}. \tag{4.31}$$

Analogous to (4.22), the inverse Langevin function $\mathcal{L}^{-1}(x)$ is replaced by the Padé approximation. Equation (4.30) involves the rubber modulus $C_R$ and the limit stretch $\lambda_L$ as material parameters.

In accordance with experimental observations on shear yielding in glassy polymers, the inelastic flow is modelled incompressible, which is given by

$$\det \mathbf{F}^p = 1. \tag{4.32}$$

Furthermore, the inelastic flow in the intermediate configuration is assumed to be irrotational, i.e. $\hat{\mathbf{l}}^p \equiv \hat{\mathbf{d}}^p$, which yields the update for the inelastic deformation as

$$\dot{\mathbf{F}}^p = \hat{\mathbf{d}}^p \mathbf{F}^p\,. \tag{4.33}$$

To comply with the inelastic incompressibility condition in equation (4.32), $\hat{\mathbf{d}}^p$ needs to be deviatoric and hence the flow rule is constitutively prescribed as

$$\hat{\mathbf{d}}^p = \dot{\gamma}^p \frac{\hat{\Sigma}^{*\prime}}{\|\hat{\Sigma}^{*\prime}\|}, \tag{4.34}$$

which is a function of the deviatoric driving stress

$$\hat{\Sigma}^{*\prime} = \hat{\Sigma}' - \hat{\boldsymbol{\tau}}^{b\prime}\,. \tag{4.35}$$

The inelastic shear strain rate in (4.34) is modelled by an Eyring-type flow

$$\dot{\gamma}^p = \dot{\gamma}_0^p \left( \exp\left[ \frac{A}{T} \left( \|\hat{\Sigma}^{*\prime}\| - s \right) \right] - \exp\left[ -\frac{A}{T} s \right] \right), \tag{4.36}$$

where $\dot{\gamma}_0^p$ and $A$ are material parameters and $T$ is the absolute temperature. Note, (4.36) is slightly different to the original double kink model by Argon [5] by taking the exponent 5/6 as 1 and by incorporating a second term which ensures $\dot{\gamma}^p = 0$ for $\|\hat{\Sigma}^{*\prime}\| = 0$. As suggested by Boyce et al. [17], the pressure dependence of inelastic flow and the softening upon yielding, from an initial value $s_0$ to a saturation value $s_s$, observed in polymers is incorporated via the yield strength

$$s(\gamma^p) = s_s + (s_0 - s_s) \exp\left[ -\frac{h\gamma^p}{s_s} \right] - \frac{\alpha_p}{3} \operatorname{tr} \hat{\Sigma}. \tag{4.37}$$

The model is implemented as a user material subroutine in the finite element programme Abaqus/Explicit[2], which requires due to the corotational framework the corotated Cauchy stress $\boldsymbol{\sigma}_r$ as update. This is readily obtained by a push-forward operation of Equation 4.29, defined as

$$\boldsymbol{\sigma}_r = \frac{1}{J} \mathbf{R}^{\mathrm{T}} \mathbf{F}^{e\,-\mathrm{T}} \hat{\Sigma} \mathbf{F}^{e\,T} \mathbf{R}, \tag{4.38}$$

where $\mathbf{R}$ is the rotation tensor obtained from the polar decomposition of $\mathbf{F}$.

The model incorporates 10 independent material parameters. We adopt the material parameters for polycarbonate (PC) at room temperature as characterised in [42] and summarised in Table 4.B.1.

**Table 4.B.1.:** Material parameters of the bulk model, being representative of PC at room temperature.

| $\mu_b/s_0$ | $\nu_b$ | $s_s/s_0$ | $As_0/T$ | $h/s_0$ | $\alpha_p$ | $\lambda_L$ | $C_R/s_0$ | $s_0$ [MPa] | $\dot{\gamma}_0^p$ [$s^{-1}$] |
|---|---|---|---|---|---|---|---|---|---|
| 11.3 | 0.38 | 0.6 | 152 | 0.06 | 1.64 | 1.64 | 0.22 | 82 | $10^{14}$ |

## 4.C. Interaction between crazing and shear yielding for material 3

In craze material 3 (cf. Table 4.1), the limit stretch $\lambda_L$ is reduced to align with the MD results, which constraints the overall axial extension of the craze fibril, leading to an essentially quasi-stiff fibril deformation behaviour. Compared to material 2, this changes under uniaxial deformation primarily the curvature of the unloading-reloading curve during cyclic loading (cf. Figure 4.4(c)). The evolution of the craze length (black line) and the crack length (magenta line) with the normalised load programme (green dashed line) for material 3 is shown in Figure 4.C.7. Similar to before, the craze length grows over multiple cycles prior to crack initiation. Thereafter, the crack propagates in each cycle. Qualitatively, the results are very similar to Figure 4.12(a). The key difference is the earlier onset of crack initiation in the 3rd cycle and the faster crack growth due to the stiffer craze fibril.

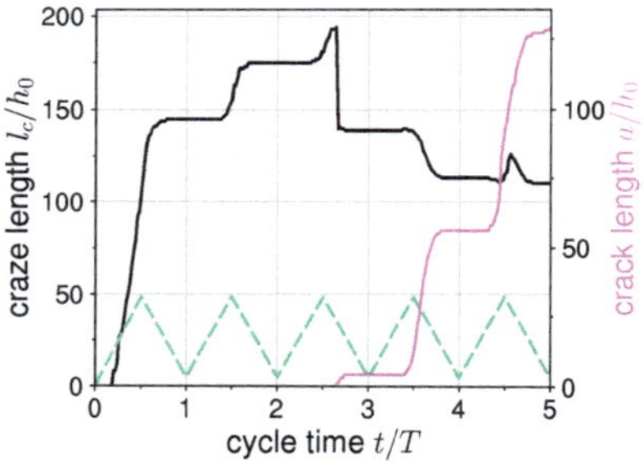


**Figure 4.C.7.:** Craze length (black) and crack length (magenta) evolution with normalised load (green) for craze material 3 in Table 4.1.

# 5. Conclusion

## 5.1. Summary

In this dissertation a micromechanics-inspired constitutive model for crazing was developed and enhanced with results from molecular dynamics simulations. The model was then employed as a traction separation law along the ligament of a mode I crack propagation problem to analyse the interaction between crazing and the adjacent bulk material in glassy polymers under cyclic loading. While more thorough summaries are given at the end of each chapter, the following essential contributions from this thesis are summed up as:

A novel micromechanical-continuum model in a finite strain framework was developed to realistically capture the craze response under cyclic loading. Derived through rigorous continuum-micromechanical considerations and based on polymer physical understanding, the model incorporates several key features. These include, for instance, (i) craze thickening due to viscoelastic deformation of existing craze fibrils, (ii) the transient process of viscoplastic conversion of intact bulk material into new fibrillated craze matter and (iii) premature jamming of the craze fibrils during unloading. The structural behaviour of craze fibrils is considered string-like, which leads to creep recovery of the fibril deformation when they are stress-free. This along with the two viscosities enables the modelling of continuous fibril drawing across loading cycles. The craze is taken to break down at a critical thickness so that the current craze fibril length can be considered a measure of cyclic damage accumulation. The micromechanical setting offers access to various physical quantities, such as the fibril stress, which could be incorporated, e.g., in other failure criteria as new knowledge emerges. Furthermore, it allows for the inclusion of measurable craze properties, such as the primordial thickness and the extension ratio, thereby simplifying material parameter calibration. Material parameters were normalised to facilitate model evaluation through a parameter study focusing on monotonic and cyclic uniaxial deformation. An essential finding of this study was the complex interplay between drawing viscosity and fibril creep viscosity. Initially, when craze fibrils are small, the drawing viscosity dominates the macroscopic response. However, as craze fibril length increases, the mechanisms shift and fibril creep viscosity exerts a more pronounced influence, particularly on the failure cycle.

A more fundamental study of polymer physics through molecular dynamics simulations was undertaken to enhance the understanding of the crazing process under cyclic loading and to address two significant uncertainties in the previous study and the model. The primary uncertainty stemmed from an incomplete understanding of the crazing mechanisms under cyclic loading. This included the structural behaviour of craze fibrils, which affects the physical picture of cyclic damage accumulation in crazes. The second uncertainty arose from the lack of experimental data, posing a twofold challenge by hindering the overall model validation and by complicating the calibration of material parameters. These issues were addressed by using molecular dynamics simulations of a flexible generic bead-spring model to study the mechanical response of sole fibrillated craze matter and the bulk-craze interaction in glassy polymers under cyclic uniaxial deformation. A salient finding was the complex stress-strain hysteresis which depends on the unloading magnitude, but was found to be qualitatively independent of the remaining bulk material and the current craze fibril length. Trough a detailed analysis, the hysteresis was attributed to a combination of constraints imposed by the entanglement network, existing pore space and pore space closure during unloading. Key findings for the continuum model included the string-like behaviour of craze fibrils during unloading, the very low craze-to-bulk stiffness and the validation that fibrils prematurely jam during unloading. The low craze stiffness highlights that the macroscopic response is governed by the craze, whereas the bulk behaves essentially as stiff. The string-like structural response of craze fibrils is a key finding, as it supports the modelled cyclic damage mechanism of fibril drawing in crazes. Based on polymer physics arguments, it was deduced that this result should qualitatively hold even in case of semiflexible bead-spring models. The additional analysis presented in Section 3.C verified this reasoning.

The essential findings from the molecular dynamics study were transferred to the continuum model by incorporating mechanisms as well as a molecular dynamics determined parameter scopes for the experimentally inaccessible material parameters. This molecular dynamics informed continuum model was then used in the finite element programme Abaqus to analyse craze and crack propagation in brittle and ductile glassy polymers under cyclic mode I loading. The parameter were normalised and a parameter scope was investigated to yield generic, but representative glassy polymer responses. The model, along with the calibration methodology, successfully reproduces important craze, shear band and failure characteristics reported in the experimental literature. These include, for instance, the craze contour, the initiation of shear bands at the current crack tip that arch towards the craze and properties of *normal fatigue crack propagation.* Particularly notable is the resemblance of the crack tip plastic zone in the simulations of ductile glassy polymers (cf. Figures 4.13 and 4.11) to the experimentally observed *epsilon-shaped deformation zone* depicted in Figure 1.2. A novel insight form this study is the delayed crack propagation due to plasticity-induced unloading of the craze at the crack tip, which required further craze growth prior to a continuation of the crack advance.

## 5.2. Outlook

The most interesting direction for future work involves carefully conducted uniaxial deformation experiments on isolated, uniformly fibrillated craze matter to study the responses under various loading conditions. The experimental setup could adopt the configuration outlined by Kambour and Kopp [47], but enhanced with modern technologies, for instance digital image correlation [91]. Utilizing a deformation controlled loading programme should allow precise control over pore space closure and enable a systematic study of the structural response of craze fibrils. This complements the molecular dynamics simulations presented in this thesis and would allow to (in-)validate the simulation results. Additionally, it would provide a helpful extension to previous experiments which mainly focused on the craze initiation stress, craze contour or crack growth rate (cf., e.g., [49]).

If the molecular dynamics results are validated, a craze fibril model could be developed based on the underlying mechanisms analysed in [57] and aiming to capture the complex craze response and its hysteresis. In a subsequent step, hysteretic dissipation could be treated as a heat source, which would allow to accommodate failure due to adiabatic fibril heating under cyclic loading. This approach could serve as the foundation for investigating the role of fibril heating in discontinuous crack growth as reported by Döll and Könczöl [23]. Depending on the outcomes, a fully coupled thermomechanical analysis could provide further insights into the competition between crazing, shear yielding and heat flow under cyclic loading without confining the analysis to the adiabatic regime. This would present a combination and continuation of Estevez et al. [28] and this work.

# Declaration of authorship

The initials Tobias Laschützta (TL), Thomas Seelig (TS), Jörg Rottler (JR) and Ting Ge (TG) are used below to mark the contributions of the respective author to the publications.

Mechanics of Materials

A continuum-micromechanical model for crazing in glassy polymers under cyclic loading

Tobias Laschuetza, Thomas Seelig

**Citation:**
Laschuetza T and Seelig T. "A continuum-micromechanical model for crazing in glassy polymers under cyclic loading". In: *Mechanics of Materials.* 189: 104901, 2024. DOI: `10.1016/j.mechmat.2023.104901`

**Declaration:**
TL is first and corresponding author. TL and TS developed the model. TL implemented and evaluated the model. TL and TS wrote the manuscript. All authors reviewed the manuscript.

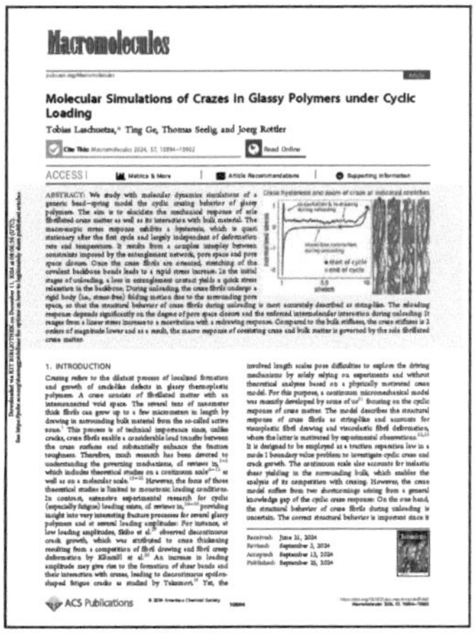
Macromolecules

Molecular Simulations of Crazes in Glassy Polymers under Cyclic Loading

Tobias Laschuetza, Ting Ge, Thomas Seelig, and Joerg Rottler

**Citation:**
Laschuetza T, Ge T, Seelig T, and Rottler J. "Molecular Simulations of Crazes in Glassy Polymers under Cyclic Loading". In: *Macromolecules.* 57(23): 10894–10902, 2024. DOI: `10.1021/acs.macromol.4c01445`

**Declaration:**
TL is first and corresponding author. TL and TS developed the research idea. JR supervised the project. TG provided the equilibrated melt configuration. TL conducted the analysis. TL and JR interpreted results. TL wrote the manuscript. All authors reviewed the manuscript.

Journal of the Mechanics and Physics of Solids

Analysis of mode I crack propagation in glassy polymers under cyclic loading using a molecular dynamics informed continuum model for crazing

Tobias Laschuetza *, Thomas Seelig

ARTICLE INFO

ABSTRACT

1. Introduction

**Citation:**
Laschuetza T and Seelig T. “Analysis of mode I crack propagation in glassy polymers under cyclic loading using a molecular dynamics informed continuum model for crazing”. In: *Journal of the Mechanics and Physics of Solids.* 194: 105901, 2025. DOI: `10.1016/j.jmps.2024.105901`

**Declaration:**
TL is first and corresponding author. TL and TS developed the research idea. TL implemented and evaluated the model in Abaqus and performed the MD calibration. TL and TS wrote the manuscript. All authors reviewed the manuscript.

# Publications and talks

## Peer-reviewed publications

- **Laschuetza T** and Seelig T. "Analysis of mode I crack propagation in glassy polymers under cyclic loading using a molecular dynamics informed continuum model for crazing". In: *Journal of the Mechanics and Physics of Solids.* 194: 105901, 2025. DOI: 10.1016/j.jmps.2024.105901.
- **Laschuetza T**, Ge T, Seelig T, and Rottler J. "Molecular Simulations of Crazes in Glassy Polymers under Cyclic Loading". In: *Macromolecules.* 57(23): 10894–10902, 2024. DOI: 10.1021/acs.macromol.4c01445.
- **Laschuetza T** and Seelig T. "A continuum-micromechanical model for crazing in glassy polymers under cyclic loading". In: *Mechanics of Materials.* 189: 104901, 2024. DOI: 10.1016/j.mechmat.2023.104901.
- Maurya MK, **Laschuetza T**, Singh MK, and Mukherji D. "Thermal Conductivity of Bottle–Brush Polymers". In: *Langmuir.* 40(8): 4392–4400, 2024. DOI: 10.1021/acs.langmuir.3c03715.
- **Laschuetza T** and Seelig T. "Remarks on dynamic cohesive fracture under static pre-stress — with a comparison to finite fracture mechanics". In: *Engineering Fracture Mechanics.* 242: 107466, 2021. DOI: 10.1016/j.engfracmech.2020.107466.

## Other publications

- **Laschuetza T** and Seelig T. "Finite element analyses of shear yielding and crazing in glassy polymers under cyclic mode I loading". In: *8th European Congress on Computational Methods in Applied Sciences and Engineering.* CIMNE, 5th - 9th Jun 2022. DOI: 10.23967/eccomas.2022.121.
- **Laschuetza T** and Seelig T. "On the energetics of dynamic cohesive crack formation". In: *Proceedings in Applied Mathematics and Mechanics.* 19(1), 2019. DOI: 10.1002/pamm.201900102.

## Conference contributions

- **Laschuetza T**, Rottler J, and Seelig T. *Competition of crazing and shear yielding under cyclic mode I loading using a molecular dynamics informed continuum micromechanical crazing model.* 6th World Congress on Computational Mechanics (WCCM) and 4th Pan American Congress on Computational Mechanics (PANACM). Vancouver, Canada, July 21–26, 2024.
- **Laschuetza T**, Ge T, Rottler J, and Seelig T. *Studying cyclic mode I crack growth with a molecular dynamics informed continuum micromechanical crazing model.* 2nd Capriccio Special Seminar (CSS). Friedrich-Alexander-Universität, Erlangen, Germany, May 16, 2024.
- **Laschuetza T**, Ge T, Seelig T, and Rottler J. *Multiscale modelling of crazing in glassy polymers under cyclic loading: molecular dynamics and continuum micromechanics.* 19th International Conference on Deformation, Yield and Fracture of Polymers (DYFP). Kerkrade, The Netherlands, Mar. 24–28, 2024.
- **Laschuetza T** and Seelig T. *Finite element analyses of shear yielding and crazing in glassy polymers under cyclic mode I loading.* 8th European Congress on Computational Methods in Applied Sciences and Engineering (ECCOMAS). Oslo, Norway, June 5–9, 2022.
- **Laschuetza T** and Seelig T. *A continuum-micromechanical model for crazing in glassy polymers under cyclic loading.* 18th International Conference on Deformation, Yield and Fracture of Polymers (DYFP). Kerkrade, The Netherlands, Apr. 10–14, 2022.
- **Laschuetza T** and Seelig T. *Computational modeling of crazing in glassy polymers under cyclic loading.* 18th European Mechanics of Materials Conference (EMMC). Oxford, England, Apr. 4–6, 2022.
- **Laschuetza T** and Seelig T. *A novel constitutive model for crazing in glassy polymers under cyclic loading.* 7th European Community on Computational Methods in Applied Sciences and Engineering (ECCOMAS) & 14th World Congress on Computational Mechanics (WCCM). Paris, France (virtual congress), Jan. 11–15, 2021.
- **Laschuetza T** and Seelig T. *Crack tip plastic zone evolution in glassy thermoplastic polymers under cyclic loading.* 29th International Workshop on Computational Mechanics of Materials (IWCMM). Dubrovnik, Croatia, Sept. 15–18, 2019.
- **Laschuetza T** and Seelig T. *On the energetics of dynamic cohesive crack formation.* 90th GAMM Annual Meeting. Vienna, Austria, Feb. 18–22, 2019.

# List of Figures

# List of Tables

# Bibliography


[1] ABAQUS. *Reference manuals*. English. United States: Dassault Systèmes Simulia Corp, 2021.

[2] ABAQUS. *Reference manuals*. English. United States: Dassault Systèmes Simulia Corp, 2023.

[3] Anand L. "On H. Hencky's Approximate Strain-Energy Function for Moderate Deformations". In: *Journal of Applied Mechanics*. 46(1): 78–82, 1979. DOI: 10.1115/1.3424532.

[4] Anand L and Gurtin ME. "A theory of amorphous solids undergoing large deformations, with application to polymeric glasses". In: *International Journal of Solids and Structures*. 40(6): 1465–1487, 2003. DOI: 10.1016/S0020-7683(02)00651-0.

[5] Argon AS. "A theory for the low-temperature plastic deformation of glassy polymers". In: *Philosophical Magazine*. 28(4): 839–865, 1973. DOI: 10.1080/14786437308220987.

[6] Argon AS and Hannoosh JG. "Initiation of crazes in polystyrene". In: *Philosophical Magazine*. 36(5): 1195–1216, 1977. DOI: 10.1080/14786437708239789.

[7] Argon AS and Salama MM. "Growth of crazes in glassy polymers". In: *Philosophical Magazine*. 36(5): 1217–1234, 1977. DOI: 10.1080/14786437708239790.

[8] Arruda EM and Boyce MC. "A three-dimensional constitutive model for the large stretch behavior of rubber elastic materials". In: *Journal of the Mechanics and Physics of Solids*. 41(2): 389–412, 1993. DOI: 10.1016/0022-5096(93)90013-6.

[9] Arruda EM, Boyce MC, and Jayachandran R. "Effects of strain rate, temperature and thermomechanical coupling on the finite strain deformation of glassy polymers". In: *Mechanics of Materials*. 19(2): 193–212, 1995. DOI: 10.1016/0167-6636(94)00034-E.

[10] Auhl R, Everaers R, Grest GS, Kremer K, and Plimpton SJ. "Equilibration of long chain polymer melts in computer simulations". In: *The Journal of Chemical Physics*. 119(24): 12718–12728, Dec. 2003. DOI: 10.1063/1.1628670.

[11] Baljon ARC and Robbins MO. "Simulations of Crazing in Polymer Glasses: Effect of Chain Length and Surface Tension". In: *Macromolecules*. 34(12): 4200–4209, 2001. DOI: 10.1021/ma0012393.

[12] Basu S, Mahajan DK, and van der Giessen E. "Micromechanics of the growth of a craze fibril in glassy polymers". In: *Polymer*. 46(18): 7504–7518, 2005.

[13] Belytschko T, Liu WK, Moran B, and Elkhodary K. *Nonlinear Finite Elements for Continua and Structures.* 2nd ed. Newark: John Wiley & Sons Inc, 2013.

[14] Bergstrom J. *Mechanics of solid polymers: Theory and computational modeling.* First edition. Amsterdam: William Andrew, 2015.

[15] Bevan L, Döll W, and Könczöl L. "Micromechanics of a craze zone at the tip of a stationary crack". In: *Journal of Polymer Science Part B: Polymer Physics.* 24(11): 2433–2444, 1986. DOI: 10.1002/polb.1986.090241103.

[16] Bevan L, Döll W, and Könczöl L. "Stress analysis of crazes at moving crack tips". In: *Polymer Bulletin.* 16(6), 1986. DOI: 10.1007/BF00255225.

[17] Boyce MC, Parks DM, and Argon AS. "Large inelastic deformation of glassy polymers. Part I: Rate dependent constitutive model". In: *Mechanics of Materials.* 7(1): 15–33, 1988. DOI: 10.1016/0167-6636(88)90003-8.

[18] Brown HR. "A molecular interpretation of the toughness of glassy polymers". In: *Macromolecules.* 24(10): 2752–2756, 1991. DOI: 10.1021/ma00010a018.

[19] Brown HR, Kramer EJ, and Bubeck RA. "Studies of craze fibril deformation during fatigue in polystyrene". In: *Journal of Polymer Science Part B: Polymer Physics.* 25(8): 1765–1778, 1987. DOI: 10.1002/polb.1987.090250816.

[20] Cohen A. "A Pade approximant to the inverse Langevin function". In: *Rheologica Acta.* 30(3): 270–273, 1991. DOI: 10.1007/BF00366640.

[21] Cotterell B. "Fracture propagation in organic glasses". In: *International Journal of Fracture Mechanics.* 4(3): 209–217, 1968. DOI: 10.1007/BF00185257.

[22] Dietz JD, Nan K, and Hoy RS. "Unexpected Ductility in Semiflexible Polymer Glasses with Entanglement Length Equal to Their Kuhn Length". In: *Physical Review Letters.* 129: 127801, 12 Sept. 2022. DOI: 10.1103/PhysRevLett.129.127801.

[23] Döll W and Könczöl L. "Micromechanics of fracture under static and fatigue loading: Optical interferometry of crack tip craze zones". In: *Crazing in Polymers Vol. 2.* Ed. by Kausch HH. Vol. 91/92. Advances in Polymer Science. Berlin, Heidelberg: Springer, Berlin, Heidelberg, 1990, pp. 137–214. DOI: 10.1007/BFb0018021.

[24] Donald AM and Kramer EJ. "The competition between shear deformation and crazing in glassy polymers". In: *Journal of Materials Science.* 17(7): 1871–1879, 1982. DOI: 10.1007/BF00540402.

[25] Ehrenstein GW. *Polymer-Werkstoffe: Struktur - Eigenschaften - Anwendung.* 3. Auflage. Hanser eLibrary. München: Hanser, 2011. DOI: 10.3139/9783446429673.

[26] Estevez R, Tijssens M, and van der Giessen E. "Modeling of the competition between shear yielding and crazing in glassy polymers". In: *Journal of the Mechanics and Physics of Solids.* 48(12): 2585–2617, 2000. DOI: 10.1016/S0022-5096(00)00016-8.

[27] Estevez R and van der Giessen E. "Modeling and Computational Analysis of Fracture of Glassy Polymers". In: *Intrinsic Molecular Mobility and Toughness of Polymers II.* Ed. by Kausch HH. Vol. 188. Advances in Polymer Science. Berlin, Heidelberg: Springer Berlin Heidelberg, 2005, pp. 195–234. DOI: 10.1007/b136977.

[28] Estevez R, Basu S, and van der Giessen E. "Analysis of temperature effects near mode I cracks in glassy polymers". In: *International Journal of Fracture Mechanics.* 132(3): 249–273, 2005. DOI: 10.1007/s10704-005-2182-1.

[29] Fett T. *Stress Intensity Factors - T-Stresses - Weight Functions.* Tech. rep. Karlsruher Institut für Technologie (KIT), 2008. 362 pp. DOI: 10.5445/KSP/1000007996.

[30] Frenkel D and Smit B. *Understanding Molecular Simulation.* Elsevier, 2002. DOI: 10.1016/B978-0-12-267351-1.X5000-7.

[31] Ge T, Grest GS, and Robbins MO. "Tensile Fracture of Welded Polymer Interfaces: Miscibility, Entanglements, and Crazing". In: *Macromolecules.* 47(19): 6982–6989, 2014. DOI: 10.1021/ma501473q.

[32] Ge T, Tzoumanekas C, Anogiannakis SD, Hoy RS, and Robbins MO. "Entanglements in Glassy Polymer Crazing: Cross-Links or Tubes?" In: *Macromolecules.* 50(1): 459–471, 2017. DOI: 10.1021/acs.macromol.6b02125.

[33] Gearing BP and Anand L. "On modeling the deformation and fracture response of glassy polymers due to shear-yielding and crazing". In: *International Journal of Solids and Structures.* 41(11): 3125–3150, 2004. DOI: 10.1016/j.ijsolstr.2004.01.017.

[34] Haward RN and Young RJ. *The Physics of Glassy Polymers.* Dordrecht: Springer Netherlands, 1997. DOI: 10.1007/978-94-011-5850-3.

[35] Helbig M, van der Giessen E, Clausen AH, and Seelig T. "Continuum-micromechanical modeling of distributed crazing in rubber-toughened polymers". In: *European Journal of Mechanics - A/Solids.* 57: 108–120, 2016. DOI: 10.1016/j.euromechsol.2015.11.007.

[36] Helbig M and Seelig T. "Micro-mechanical modeling of fibrillation in amorphous polymers". In: *Computational Materials Science.* 52(1): 118–122, 2012. DOI: 10.1016/j.commatsci.2011.04.007.

[37] Hempel P. "Constitutive modeling of amorphous thermoplastic polymers with special emphasis on manufacturing processes". PhD thesis. Karlsruher Institut für Technologie (KIT), 2016. DOI: 10.5445/KSP/1000056493.

[38] Hoare J and Hull D. "Craze yielding and stress-strain characteristics of crazes in polystyrene". In: *The Philosophical Magazine: A Journal of Theoretical Experimental and Applied Physics.* 26(2): 443–455, 1972. DOI: 10.1080/14786437208227440.

[39] Holopainen S. "Modeling of the mechanical behavior of amorphous glassy polymers under variable loadings and comparison with state-of-the-art model predictions". In: *Mechanics of Materials.* 66: 35–58, 2013. DOI: 10.1016/j.mechmat.2013.06.009.

[40] Holzapfel GA. *Nonlinear solid mechanics: A continuum approach for engineering.* Repr. Chichester and Weinheim: Wiley, 2010.

[41] Hoy RS and Robbins MO. "Effect of equilibration on primitive path analyses of entangled polymers". In: *Physical Review E.* 72: 061802, 6 Dec. 2005. DOI: 10.1103/PhysRevE.72.061802.

[42] Hund J. "Characterisation and modelling of PC/ABS blends". PhD thesis. Karlsruhe, 2022. DOI: 10.5445/KSP/1000141093.

[43] Imai Y and Ward IM. "A study of craze deformation in the fatigue fracture of polymethylmethacrylate". In: *Journal of Materials Science.* 20(11): 3842–3852, 1985. DOI: 10.1007/BF00552372.

[44] Ishikawa M and Narisawa I. "Fracture of notched polycarbonate under hydrostatic pressure". In: *Journal of Materials Science.* 18(7): 1947–1957, 1983. DOI: 10.1007/BF00554987.

[45] Ishikawa M, Narisawa I, and Ogawa H. "Criterion for craze nucleation in polycarbonate". In: *Journal of Polymer Science: Polymer Physics Edition.* 15(10): 1791–1804, 1977. DOI: 10.1002/pol.1977.180151009.

[46] Kambour RP. "A review of crazing and fracture in thermoplastics". In: *Journal of Polymer Science: Macromolecular Reviews.* 7(1): 1–154, 1973. DOI: 10.1002/pol.1973.230070101.

[47] Kambour RP and Kopp RW. "Cyclic stress–strain behavior of the dry polycarbonate craze". In: *Journal of Polymer Science Part A-2: Polymer Physics.* 7(1): 183–200, 1969. DOI: 10.1002/pol.1969.160070115.

[48] Kausch HH, ed. *Crazing in polymers.* Vol. 52/53. Advances in Polymer Science. Berlin: Springer, Berlin, Heidelberg, 1983. DOI: 10.1007/BFb0024054.

[49] Kausch HH, ed. *Crazing in Polymers Vol. 2.* Advances in Polymer Science. Berlin, Heidelberg: Springer Berlin Heidelberg, 1990. DOI: 10.1007/BFb0018017.

[50] Könczöl L, Döll W, and Bevan L. "Mechanisms and micromechanics of fatigue crack propagation in glassy thermoplastics". In: *Colloid & Polymer Science.* 268(9): 814–822, 1990. DOI: 10.1007/BF01410959.

[51] Kramer EJ and Hart EW. "Theory of slow, steady state crack growth in polymer glasses". In: *Polymer.* 25(11): 1667–1678, 1984. DOI: 10.1016/0032-3861(84)90164-2.

[52] Kramer EJ. "Microscopic and molecular fundamentals of crazing". In: *Crazing in polymers.* Ed. by Kausch HH. Vol. 52-53. Advances in Polymer Science. Berlin: Springer, 1983, pp. 1–56. DOI: 10.1007/BFb0024055.

[53] Kramer EJ and Berger LL. "Fundamental processes of craze growth and fracture". In: *Crazing in Polymers Vol. 2.* Ed. by Kausch HH. Vol. 91/92. Advances in Polymer Science. Berlin, Heidelberg: Springer, Berlin, Heidelberg, 1990, pp. 1–68. DOI: 10.1007/BFb0018018.

[54] Kremer K and Grest GS. "Dynamics of entangled linear polymer melts: A molecular-dynamics simulation". In: *The Journal of Chemical Physics.* 92(8): 5057–5086, Apr. 1990. DOI: 10.1063/1.458541.

[55] Laschuetza T, Ge T, Seelig T, and Rottler J. "Molecular Simulations of Crazes in Glassy Polymers under Cyclic Loading". In: *Macromolecules.* 57(23): 10894–10902, 2024. DOI: 10.1021/acs.macromol.4c01445.

[56] Laschuetza T and Seelig T. "A continuum-micromechanical model for crazing in glassy polymers under cyclic loading". In: *Mechanics of Materials.* 189: 104901, 2024. DOI: 10.1016/j.mechmat.2023.104901.

[57] Laschuetza T and Seelig T. "Analysis of mode I crack propagation in glassy polymers under cyclic loading using a molecular dynamics informed continuum model for crazing". In: *Journal of the Mechanics and Physics of Solids.* 194: 105901, 2025. DOI: 10.1016/j.jmps.2024.105901.

[58] Leonov AI and Brown HR. "A model of fibril deformation in crazes". In: *Journal of Polymer Science Part B: Polymer Physics.* 29(2): 197–209, 1991. DOI: 10.1002/polb.1991.090290206.

[59] Mahajan DK and Hartmaier A. "Mechanisms of crazing in glassy polymers revealed by molecular dynamics simulations". In: *Physical Review E.* 86: 021802, 2 Aug. 2012. DOI: 10.1103/PhysRevE.86.021802.

[60] Mahajan DK, Singh B, and Basu S. "Void nucleation and disentanglement in glassy amorphous polymers". In: *Physical Review E.* 82: 011803, 1 July 2010. DOI: 10.1103/PhysRevE.82.011803.

[61] Maiti S and Geubelle PH. "A cohesive model for fatigue failure of polymers". In: *Engineering Fracture Mechanics.* 72(5): 691–708, 2005. DOI: 10.1016/j.engfracmech.2004.06.005.

[62] McCrum NG, Buckley CP, and Bucknall CB. *Principles of polymer engineering.* 2. ed. Oxford: Oxford Univ. Press, 1997.

[63] Mills NJ and Walker N. "Fatigue crack initiation in glassy plastics in high strain fatigue tests". In: *Journal of Materials Science.* 15(7): 1832–1840, 1980. DOI: 10.1007/BF00550604.

[64] Mills PJ, Kramer EJ, and Brown HR. "Real time small-angle X-ray scattering from polystyrene crazes during fatigue". In: *Journal of Materials Science.* 20(12): 4413–4420, 1985. DOI: 10.1007/BF00559330.

[65] Nan K and Hoy RS. "Craze Extension Ratio of Semiflexible Polymer Glasses". In: *Macromolecules.* 56(20): 8369–8375, 2023. DOI: 10.1021/acs.macromol.3c01608.

[66] Narisawa I and Yee A. "Crazing and Fracture of Polymers". In: *Materials Science and Technology.* Vol. 12. 1993, pp. 699–765. DOI: 10.1002/9783527603978.mst0146.

[67] Oxborough RJ and Bowden PB. "A general critical-strain criterion for crazing in amorphous glassy polymers". In: *Philosophical Magazine.* 28(3): 547–559, 1973. DOI: 10.1080/14786437308221002.

[68] Paszke A et al. "PyTorch: An Imperative Style, High-Performance Deep Learning Library". In: *Proceedings of the 33rd International Conference on Neural Information Processing Systems*. Red Hook, NY, USA: Curran Associates Inc, 2019, pp. 8026–8037.

[69] Plimpton S. "Fast Parallel Algorithms for Short-Range Molecular Dynamics". In: *Journal of Computational Physics*. 117(1): 1–19, 1995. DOI: 10.1006/jcph.1995.1039.

[70] Pruitt L and Suresh S. "Cyclic stress fields for fatigue cracks in amorphous solids Experimental measurements and their implications". In: *Philosophical Magazine A*. 67(5): 1219–1245, 1993. DOI: 10.1080/01418619308224768.

[71] Pulos GC and Knauss WG. "Nonsteady crack and craze behavior in PMMA under cyclical loading: III. Effect of load history on cohesive force distribution on the craze". In: *International Journal of Fracture*. 93(1): 187–207, 1998. DOI: 10.1023/A:1007569703250.

[72] Rabinowitz S and Beardmore P. "Cyclic deformation and fracture of polymers". In: *Journal of Materials Science*. 9(1): 81–99, 1974. DOI: 10.1007/BF00554758.

[73] Rice JR. *Mechanics of Crack Tip Deformation and Extension by Fatigue*. Fatigue Crack Propagation. ASTM International, 1967, pp. 247–311.

[74] Rottler J, Barsky S, and Robbins MO. "Cracks and Crazes: On Calculating the Macroscopic Fracture Energy of Glassy Polymers from Molecular Simulations". In: *Physical Review Letters*. 89: 148304, 14 Sept. 2002. DOI: 10.1103/PhysRevLett.89.148304.

[75] Rottler J and Robbins MO. "Jamming under Tension in Polymer Crazes". In: *Physical Review Letters*. 89: 195501, 19 Oct. 2002. DOI: 10.1103/PhysRevLett.89.195501.

[76] Rottler J and Robbins MO. "Growth, microstructure, and failure of crazes in glassy polymers". In: *Physical Review E*. 68: 011801, 1 July 2003. DOI: 10.1103/PhysRevE.68.011801.

[77] Sauer JA and Hara M. "Effect of molecular variables on crazing and fatigue of polymers". In: *Crazing in Polymers Vol. 2*. Ed. by Kausch HH. Advances in Polymer Science. Berlin, Heidelberg: Springer, Berlin, Heidelberg, 1990, pp. 69–118. DOI: 10.1007/BFb0018019.

[78] Schirrer R. "Optical interferometry: Running crack-tip morphologies and craze material properties". In: *Crazing in Polymers Vol. 2*. Ed. by Kausch HH. Vol. 91/92. Berlin, Heidelberg: Springer, Berlin, Heidelberg, 1990, pp. 215–261. DOI: 10.1007/BFb0018022.

[79] Schirrer R, Le Masson J, Tomatis B, and Lang R. "The disentanglement time of the craze fibrils under cyclic loading". In: *Polymer Engineering and Science*. 24(10): 820–824, 1984. DOI: 10.1002/pen.760241012.

[80] Seelig T. "Computational modeling of deformation mechanisms and failure in thermoplastic multilayer composites". In: *Composites Science and Technology*. 68(5): 1198–1208, 2008. DOI: 10.1016/j.compscitech.2007.07.017.

[81] Sha Y, Hui CY, Ruina A, and Kramer EJ. "Continuum and Discrete Modeling of Craze Failure at a Crack Tip in a Glassy Polymer". In: *Macromolecules.* 28(7): 2450–2459, 1995. DOI: 10.1021/ma00111a044.

[82] Sharma R, Boyce MC, and Socrate S. "Micromechanics of toughening in ductile/brittle polymeric microlaminates: Effect of volume fraction". In: *International Journal of Solids and Structures.* 45(7-8): 2173–2202, 2008. DOI: 10.1016/j.ijsolstr.2007.11.023.

[83] Skibo MD, Hertzberg RW, Manson JA, and Kim SL. "On the generality of discontinuous fatigue crack growth in glassy polymers". In: *Journal of Materials Science.* 12(3): 531–542, 1977. DOI: 10.1007/BF00540278.

[84] Socrate S, Boyce MC, and Lazzeri A. "A micromechanical model for multiple crazing in high impact polystyrene". In: *Mechanics of Materials.* 33(3): 155–175, 2001. DOI: 10.1016/S0167-6636(00)00068-5.

[85] Souza Neto EA de, Perić D, and Owen DRJ. *Computational Methods for Plasticity.* Wiley, 2008. DOI: 10.1002/9780470694626.

[86] Sternstein SS and Myers FA. "Yielding of glassy polymers in the second quadrant of principal stress space". In: *Journal of Macromolecular Science, Part B.* 8(3-4): 539–571, 1973. DOI: 10.1080/00222347308201636.

[87] Sternstein SS and Ongchin L. "Yield criteria for plastic deformation on glassy high polymers in general stress fields". In: *Polymer Preprints.* 10: 1117–1124, 1969.

[88] Strobl G. *The Physics of Polymers.* Berlin, Heidelberg: Springer Berlin Heidelberg, 2007. DOI: 10.1007/978-3-540-68411-4.

[89] Stukowski A. "Visualization and analysis of atomistic simulation data with OVITO–the Open Visualization Tool". In: *Modelling and Simulation in Materials Science and Engineering.* 18(1): 015012, Dec. 2009. DOI: 10.1088/0965-0393/18/1/015012.

[90] Suresh S. *Fatigue of Materials.* 2nd ed. Cambridge University Press, 1998.

[91] Sutton MA, Orteu JJ, and Schreier H. *Image Correlation for Shape, Motion and Deformation Measurements.* Boston, MA: Springer US, 2009. DOI: 10.1007/978-0-387-78747-3.

[92] Takemori MT and Kambour RP. "Discontinuous fatigue crack growth in polycarbonate". In: *Journal of Materials Science.* 16(4): 1108–1110, 1981. DOI: 10.1007/BF00542759.

[93] Takemori MT. "Fatigue fracture of polycarbonate". In: *Polymer Engineering and Science.* 22(15): 937–945, 1982. DOI: 10.1002/pen.760221506.

[94] Takemori MT. "Competition between crazing and shear flow during fatigue". In: *Crazing in Polymers Vol. 2.* Ed. by Kausch HH. Vol. 91/92. Advances in Polymer Science. Berlin, Heidelberg: Springer, Berlin, Heidelberg, 1990, pp. 263–300. DOI: 10.1007/BFb0018023.

[95] Thompson AP et al. "LAMMPS - a flexible simulation tool for particle-based materials modeling at the atomic, meso, and continuum scales". In: *Computer Physics Communications.* 271: 108171, 2022. DOI: 10.1016/j.cpc.2021.108171.

[96] Tijssens M, Giessen Ed, and Sluys LJ. "Simulation of mode I crack growth in polymers by crazing". In: *International Journal of Solids and Structures.* 37(48-50): 7307–7327, 2000. DOI: 10.1016/S0020-7683(00)00200-6.

[97] Tijssens M, van der Giessen E, and Sluys LJ. "Modeling of crazing using a cohesive surface methodology". In: *Mechanics of Materials.* 32(1): 19–35, 2000. DOI: 10.1016/S0167-6636(99)00044-7.

[98] Toepperwein GN and Pablo JJ de. "Cavitation and Crazing in Rod-Containing Nanocomposites". In: *Macromolecules.* 44(13): 5498–5509, 2011. DOI: 10.1021/ma200541s.

[99] Ungsuwarungsri T and Knauss WG. "A Nonlinear Analysis of an Equilibrium Craze: Part I—Problem Formulation and Solution". In: *Journal of Applied Mechanics.* 55(1): 44–51, 1988. DOI: 10.1115/1.3173659.

[100] Venkatesan S and Basu S. "Investigations into crazing in glassy amorphous polymers through molecular dynamics simulations". In: *Journal of the Mechanics and Physics of Solids.* 77(46): 123–145, 2015. DOI: 10.1016/j.jmps.2015.01.005.

[101] Virtanen P et al. "SciPy 1.0: Fundamental Algorithms for Scientific Computing in Python". In: *Nature Methods.* 17: 261–272, 2020. DOI: 10.1038/s41592-019-0686-2.

[102] Wang J, 't Veld PJ in, Robbins MO, and Ge T. "Effects of Coarse-Graining on Molecular Simulation of Craze Formation in Polymer Glass". In: *Macromolecules.* 55(4): 1267–1278, 2022. DOI: 10.1021/acs.macromol.1c01969.

[103] Wang WCV and Kramer EJ. "A distributed dislocation stress analysis for crazes and plastic zones at crack tips". In: *Journal of Materials Science.* 17(7): 2013–2026, 1982. DOI: 10.1007/BF00540419.

[104] Warren WE, Chudnovsky A, and Mullen RL. "On the accuracy of the calculated stress field around a craze". In: *Polymer Engineering and Science.* 29(6): 426–431, 1989. DOI: 10.1002/pen.760290612.

[105] Wu JBC and Li JCM. "Slip processes in the deformation of polystyrene". In: *Journal of Materials Science.* 11(3): 434–444, 1976. DOI: 10.1007/BF00540924.

[106] Wu PD and van der Giessen E. "On improved network models for rubber elasticity and their applications to orientation hardening in glassy polymers". In: *Journal of the Mechanics and Physics of Solids.* 41(3): 427–456, 1993. DOI: 10.1016/0022-5096(93)90043-F.

[107] Yu W and Blair M. "DNAD, a simple tool for automatic differentiation of Fortran codes using dual numbers". In: *Computer Physics Communications.* 184(5): 1446–1452, 2013. DOI: 10.1016/j.cpc.2012.12.025.

## Schriftenreihe des Instituts für Mechanik
## ISSN 2363-4936

Hrsg. von: Prof. Dr.-Ing. habil. Peter Betsch, Prof. Dr.-Ing. habil. Thomas Seelig

---

Band 1 Marlon Franke
**Discretisation techniques for large deformation computational contact elastodynamics.**
ISBN 978-3-7315-0278-4

Band 2 Philipp Hempel
**Constitutive modeling of amorphous thermoplastic polymers with special emphasis on manufacturing processes.**
ISBN 978-3-7315-0550-1

Band 3 Martin Helbig
**Mehrskalenmodellierung von Schädigung in gummimodifizierten thermoplastischen Kunststoffen.**
ISBN 978-3-7315-0571-6

Band 4 Maik Dittmann
**Isogeometric analysis and hierarchical refinement for multi-field contact problems.**
ISBN 978-3-7315-0616-4

Band 5 Yinping Yang
**Numerical methods for the inverse dynamics simulation of underactuated mechanical systems.**
ISBN 978-3-7315-0626-3

Band 6 Alexander Janz
**Structure-preserving space-time discretization in a mixed framework for multi-field problems in large strain elasticity.**
ISBN 978-3-7315-0926-4

Band 7 Mark Georg Schiebl
**Thermodynamically consistent space-time discretization of non-isothermal mechanical systems in the framework of GENERIC.**
ISBN 978-3-7315-1117-5

Band 8 Jonas Hund
**Characterisation and modelling of PC/ABS blends.**
ISBN 978-3-7315-1157-1

---

Die Bände sind unter www.ksp.kit.edu als PDF frei verfügbar oder als Druckausgabe bestellbar.

Band 9 Jonathan Schulte
**Multi-field modeling and simulation of fiber-reinforced polymers.**
ISBN 978-3-7315-1251-6

Band 10 Robin Pfefferkorn
**EAS Elements for Solid Mechanics – Mesh Distortion Insensitive and Hourglassing-Free Formulations with Increased Robustness.**
ISBN 978-3-7315-1271-4

Band 11 Timo Ströhle
**Inverse dynamics of underactuated flexible mechanical systems governed by quasi-linear hyperbolic partial differential equations.**
ISBN 978-3-7315-1336-0

Band 12 Tobias Laschütza
**Multiscale modelling of crazing in glassy polymers under cyclic loading.**
ISBN 978-3-7315-1409-1

---